C000171554

HIGHER ALGEBRA FOR SCHOOLS

BY

W. L. FERRAR, M.A.
FELLOW OF HERTFORD COLLEGE
OXFORD

OXFORD
AT THE CLARENDON PRESS

Oxford University Press, Amen House, London E.C.4
GLASGOW NEW YORK TORONTO MELBOURNE WELLINGTON
BOMBAY CALCUTTA MADRAS KARACHI KUALA LUMPUR
CAPE TOWN IBADAN NAIROBI ACCRA

FIRST PUBLISHED 1945
REPRINTED LITHOGRAPHICALLY IN GREAT BRITAIN
AT THE UNIVERSITY PRESS, OXFORD
FROM SHEETS OF THE FIRST EDITION
1948, 1952, 1956, 1959

PREFACE

THIS book is written primarily for two classes of readers:

(1) young mathematicians who have passed the stage roughly indicated by 'School Certificate with additional mathematics', and who are just beginning their more advanced work;

(2) young scientists, at the same stage, who are beginning a composite course in mathematics and science.

The book has been carefully divided into starred and unstarred sections; all sections marked with a star (asterisk) are intended primarily for the mathematicians and not for the scientists. The latter should omit all starred sections, at any rate on a first reading.

The explanations given in the text are sufficiently full for the book to be used either by a mathematical set working under a teacher or by individuals reading very much on their own with only occasional help. I hope that, even where the teacher normally counts on explaining matters himself and using the book as a source of examples and supplementary explanation, he will, from time to time, abstain from blackboard work and tell the class to study a part of the book for themselves. In the last forty years the standard of teaching has so vastly improved that there is often no need for the class to learn from a book; they have merely to wait and be taught. This is excellent so far as it goes, but it fails to develop the power of learning for oneself, a power which, later on, is at least as important as the knowledge one has gained. And so I hope that even the most expert teacher will sometimes allow his class to battle with their own difficulties before he comes to their aid.

Of the subject-matter little need be said in a preface. It must stand or fall on its own merits. My chief concern has been to present an ordered development of this part of Algebra, and to provide both scientist and mathematician with the knowledge they will need later on.

The treatment, given in Chapters X and XI, of the binomial, exponential, and logarithmic series is intended as an introduction that defers all the logical difficulties. Many students will need to be familiar with these series and to have acquired the first rough idea of the representation of a function by a non-terminating series: comparatively few will ever need much more. My own teaching experience is that attempts to put the logical difficulties too early result in confusion of mind.

I have tried so to treat the subject that anyone who has studied my book will be able to face a Higher Certificate examination, or a School Leaving examination, without much special preparation in algebra. This does not mean that I have disregarded the details of the examination altogether. I have been in close touch with the recent discussions concerning the reform of examination syllabuses and, in particular, with the work of the Cambridge Joint Advisory Committee for Mathematics. In fact, the book conforms fairly closely to the syllabus recommended by that committee, ending at the point where the young scientist and young mathematician part company.

I hope to publish a sequel to the present volume. This will be for the mathematician only and will deal with topics not recommended for study at school by the scientist.

In conclusion, I offer my sincere thanks to the staff of the Oxford University Press for the excellence of their work, an excellence that seems to have resisted every difficulty that war has brought against it.

W. L. F.

HERTFORD COLLEGE
OXFORD
21 October 1944

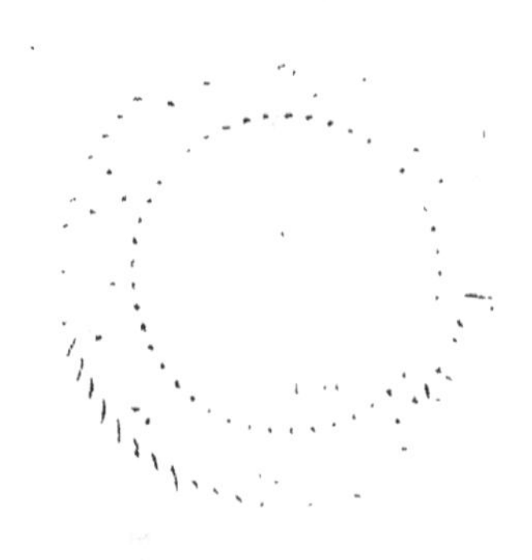

CONTENTS

A SUMMARY OF ELEMENTARY ALGEBRA

THE reader should know the items in this list before he starts to read the book.

Factors and easy expansions

$$x^2-y^2 = (x-y)(x+y), \qquad (x+y)^2 = x^2+2xy+y^2,$$
$$x^3-y^3 = (x-y)(x^2+xy+y^2), \qquad x^3+y^3 = (x+y)(x^2-xy+y^2),$$
$$(x+y)^3 = x^3+3x^2y+3xy^2+y^3,$$
$$(x+y+z)^2 = x^2+y^2+z^2+2yz+2zx+2xy,$$
$$x^3+y^3+z^3-3xyz = (x+y+z)(x^2+y^2+z^2-yz-zx-xy).$$

All these, if they are not known already, can be verified by actual multiplication.

Solution of simple equations, including simultaneous equations:

(a) two linear equations,

(b) one linear and one quadratic equation.

Simple ratio and proportion, including

(i) 'if $\frac{a}{b} = \frac{c}{d}$, then $ad = bc$',

(ii) 'if $\frac{a}{b} = \frac{c}{d}$, and each ratio $= k$, then $a = bk$, $c = dk$, and when $k \neq 1$,
$$\frac{a+b}{a-b} = \frac{c+d}{c-d}\text{'}.$$

The latter, if it is not known already, can be verified by substituting bk for a and dk for c.

Graphs. The method of plotting points and joining them to form a continuous graph.

Solution of quadratic equations

The roots of the equation
$$ax^2+bx+c = 0$$
are given by the formula
$$\frac{-b\pm\sqrt{(b^2-4ac)}}{2a}.$$

This formula is found by 'completing the square'. Solution by formula is a bad method in the early stages of algebra, but it is useful once the early stages have been passed. The formula is proved in Chapter III.

Indices

$$a^{-p} = \frac{1}{a^p}, \qquad a^{p/q} = \sqrt[q]{a^p}, \qquad a^0 = 1;$$

$$a^m \times a^n = a^{m+n}, \qquad a^m \div a^n = a^{m-n}, \qquad (a^m)^n = a^{mn}.$$

Logarithms

$$\log_a(xy) = \log_a x + \log_a y; \ \log_a(x^n) = n \log_a x.$$

The proofs of these formulae are given in Chapter XI.

Progressions

$a, a+d, a+2d, \ldots$ is an arithmetic progression;

the arithmetic mean of a and b is $\frac{1}{2}(a+b)$.

$a, ar, ar^2, \ldots$ is a geometric progression;

the geometric mean of a and b is $\sqrt{(ab)}$.

No further knowledge of progressions is presumed. It is, however, useful to know the formula

$$\frac{1-r^n}{1-r} = 1+r+r^2+\ldots+r^{n-1},$$

which can be proved either by carrying out the long division of $1-r^n$ by $1-r$, or by multiplying the right-hand side by $1-r$.

It is also useful to note that, in many problems, the simplest notation for three numbers in arithmetic progression is

$$a-d, a, a+d$$

and for four numbers in arithmetic progression is

$$a-3d, a-d, a+d, a+3d.$$

READING FOR SCIENTISTS

OMIT all starred examples.

Chapter I. Omit § 2.4.

Chapter II. Do not attempt to memorize the proofs of § 4; note the worked examples in § 5. Omit § 7.

Chapter III.

Chapter IV. Read § 1, but omit the rest.

Chapter V. Omit § 5.

Chapter VI. Omit § 7.

Chapter VII. Omit § 4.

Chapter VIII. Omit § 4.

Chapters X and XI.

Chapter XIII. Omit § 6 and all after § 7.3.

This is a minimum course; it may be added to if time permits. In particular, Chapter XII should be studied, if possible.

CHAPTER I

CONSTANTS; VARIABLES; FUNCTIONS

1. Constants and variables

From the earliest stages of algebra we are accustomed to use letters, say x, y, v, or a, to denote numbers. In this use of letters to denote numbers there is one point that calls for explicit mention at the outset. This point is the distinction between constants and variables.

When we consider an expression such as $6x^2-3x+4$ for different values of x, say, for example, when we plot the values of $6x^2-3x+4$ for values of x between $x = -4$ and $x = 3$, the letter x is thought of as taking more than one value in the course of the work, or, in the older phrase, x is considered 'to vary'. Such a letter is called a variable. On the other hand, each of the numbers 6, -3, 4 is fixed and unchanging, or, in the older phrase, is 'constant'. Such numbers are commonly referred to as constants. In many problems we must consider the values of expressions such as ax^2+bx+c when the letters a, b, c denote fixed (or constant) numbers, which do not change, while the value of x changes (or varies). In dealing with such a problem we refer to a, b, c as constants: and we refer to x as a variable.

DEFINITION 1. *A letter that denotes one fixed number, and that number only, is called a* CONSTANT: *a letter that is not restricted to one value, but may take different values, is called a* VARIABLE.

2. Functions of a variable

2.1. Let x denote a variable. Let y denote a second variable related to x in such a way that the value of y is fixed once the value of x has been fixed. Then y is said to be a function of x. For example, if y is defined in terms of x by any one of the formulae

$$y = 2x+3, \qquad y = \frac{3x^2+4x+5}{5x^2+4x+3}, \qquad y = x^x,$$

the precise value of y is fixed once the value of x has been fixed.

Let us suppose, for instance, that x is 2: then the three formulae fix the value of y as 7, $\frac{25}{31}$, 4 respectively.

In the study of algebra we shall be chiefly concerned with the simpler types of functions, such as polynomials, rational functions, and, at a later stage, what are called algebraic functions. We begin by defining these types of function.

2.2. In the expression $2x+3$ the variable x appears only to the first power: in the expression $4x^2+2x+3$ the variable x appears only to the first and second powers, there being no term x^3, x^4, or higher power. It is convenient to have distinguishing names to mark the highest power of x that appears. Accordingly, we note the following definitions:

DEFINITION 2. *When a, b, c, d denote constants, and x is a variable, the functions*

$$ax+b, \qquad ax^2+bx+c, \qquad ax^3+bx^2+cx+d \quad (a \neq 0),$$

are called, respectively, LINEAR, QUADRATIC, CUBIC *functions of x.*

DEFINITION 3. *When a,..., k denote constants, and x is a variable, the function*

$$ax^n+bx^{n-1}+\ldots+jx+k \quad (a \neq 0), \tag{1}$$

wherein n is a positive integer, is called a POLYNOMIAL FUNCTION *of x of degree n; sometimes it is called simply a* POLYNOMIAL. *The constants a, b,..., j, k are referred to as the* COEFFICIENTS *of x^n, x^{n-1},..., x, x^0 respectively; the separate parts of the sum (1), namely, ax^n, bx^{n-1},..., are referred to as the* TERMS *of the polynomial.*

There are many phrases, such as 'a polynomial of degree n in x', or 'a polynomial in x of the nth degree', that serve equally well to denote the function (1) above. The exact form of words is of little moment.

It will be noted that Definition 2 gives particular names to the polynomial when n has the value 1, 2, or 3. This is purely a matter of convenience; and it is sometimes convenient to use the particular names QUARTIC (or BIQUADRATIC), QUINTIC, SEXTIC for polynomials of degree 4, 5, 6 respectively.

2.3. DEFINITION 4. *When $P(x)$ and $Q(x)$ denote two distinct polynomial functions of x, the function*

$$\frac{P(x)}{Q(x)},$$

which denotes the fraction '$P(x)$ divided by $Q(x)$', is called a RATIONAL FUNCTION *of x.*

For example,

$$\frac{2x+3}{3x^2+8x+1}, \qquad \frac{9x^3+16}{14x+3}, \qquad \frac{7x^{10}+13}{6x^{10}+1}$$

are rational functions of the variable x. Each function is the 'ratio' of one polynomial to another polynomial, and this is the origin of the term 'rational function'.

For convenience in printing the fraction '$P(x)$ divided by $Q(x)$' is often denoted by $P(x)/Q(x)$. This notation is not recommended for use in written work, where it sometimes leads to mistakes. In print the notation saves space and is easier for the compositor to set up.

2.4.* There are other types of function that we shall occasionally encounter; in particular, ALGEBRAIC FUNCTIONS. We shall not elaborate their general study, but the reader should endeavour to obtain a clear idea of their type. We illustrate the points at issue by an example.

Suppose that, in the study of some problem, we have two variables x and y and that we have been able to prove the relation

$$y^2-(2x+1)y+x^2 = 0. \tag{2}$$

This equation is a quadratic equation in y; it is $y^2+Ay+B = 0$, where A and B are polynomials in x. In such a case we refer to y as an algebraic function of x. By solving the quadratic equation we see that

$$y = \frac{2x+1+\sqrt{(4x+1)}}{2} \tag{3}$$

gives an explicit formula for one of the possible values of y if (2) is to be satisfied.

In general, if there is a relation of the form

$$y^n+P_1(x)y^{n-1}+P_2(x)y^{n-2}+...+P_n(x) = 0,$$

where $P_1(x),\ldots, P_n(x)$ are rational functions of x, the variable y is said to be an algebraic function of x. In the general case, when $n > 4$, there may or may not be an explicit formula† that corresponds to (3).

In this book we shall be concerned only with particular cases of algebraic functions and not at all with their general theory.

3. Dependent and independent variables

When a variable y is defined in terms of a variable x, we may, when convenient, distinguish between them by using the titles 'dependent variable' and 'independent variable'. Thus, when we are considering

$$y = 3x^2+4x-3,$$

and we think of x as taking any value it pleases, we call x the INDEPENDENT VARIABLE: the value of y is fixed only when the value of x is fixed and so we call y the DEPENDENT VARIABLE.

EXAMPLES I A

1. State whether the following functions are polynomial or rational functions of x, and, if they are polynomial, state whether they are linear, quadratic, or cubic polynomials:

$$3x+4, \quad \frac{2x+3}{9x^2+7}, \quad 6x^3+27x, \quad \frac{9x^2+7}{2x+3}, \quad \frac{1}{3x},$$

2. Prove that the sum and product of the two polynomials

$$2x^2+3, \quad 3x^3+2x^2+5x+7$$

are polynomials of degree 3 and 5 respectively.

3.* Prove that the sum of two polynomials, of degrees m and n, where $m \geqslant n$, is, in general, a polynomial of degree m; and that the product is always a polynomial of degree $m+n$.

4. Prove that the sum and product of the two rational functions

$$\frac{2x^2+3}{4x+1}, \quad \frac{3x^2-1}{x^3-2x-1}$$

are also rational functions. Prove, further, that the result of dividing the first function by the second is a rational function.

† It is a theorem of higher algebra that no series of root extractions, such as give the roots of a quadratic or cubic equation, can suffice to furnish an explicit formula for the roots of the general equation of degree 5 or more.

SOLUTION. *Sum*

$$\frac{2x^2+3}{4x+1}+\frac{3x^2-1}{x^3-2x-1} = \frac{(2x^2+3)(x^3-2x-1)+(3x^2-1)(4x+1)}{(4x+1)(x^3-2x-1)}, \quad (1)$$

which is a polynomial of degree 5 in x divided by a polynomial of degree 4 in x, and is therefore a rational function of x. [NOTE. There is no interest in the calculation of the actual coefficients in the numerator and denominator: it is the general form of the expression (1) with which we are concerned.]

5.* Given two rational functions of x, prove that their sum, their product, and the quotient of either by the other, are also rational functions of x.

6.* Prove that when

$$y = \frac{\sqrt{(1+x)}+\sqrt{x}}{\sqrt{(1+x)}-\sqrt{x}},$$

y is an algebraic function of x satisfying the equation

$$y^2-2y(1+2x)+1 = 0.$$

SOLUTION. We shall outline two methods, both useful examples of a general line of attack on this and similar problems.

(i) When

$$\frac{y}{1} = \frac{\sqrt{(1+x)}+\sqrt{x}}{\sqrt{(1+x)}-\sqrt{x}}, \quad (1)$$

$$\frac{y+1}{y-1} = \frac{\sqrt{(1+x)}}{\sqrt{x}},$$

and so

$$x(y+1)^2 = (1+x)(y-1)^2.$$

Therefore

$$y^2-2y(1+2x)+1 = 0,$$

and y is an algebraic function of x.

(ii) Write $x = u^2$, $1+x = v^2$ (to dispense with root signs). Then

$$y = \frac{v+u}{v-u},$$

i.e.

$$y(v-u) = v+u,$$

i.e.

$$v(y-1) = u(y+1).$$

Hence

$$v = u\frac{y+1}{y-1}.$$

But, by the definitions of u and v, $1+u^2 = v^2$, so that

$$1+u^2 = u^2\frac{(y+1)^2}{(y-1)^2}.$$

On writing $u^2 = x$ and clearing of fractions,

$$(y-1)^2(1+x) = x(y+1)^2,$$

as in (i).

7.* Prove that, when

$$y = \frac{\sqrt[3]{(1+x)}+\sqrt{x}}{\sqrt[3]{(1+x)}+2\sqrt{x}},$$

y is an algebraic function of x satisfying the equation

$$x^3(2y-1)^6 = (1+x)^2(1-y)^6.$$

HINT. Use the methods of Example 6.

8.* (*Harder.*) Prove that, when

$$y = \frac{\sqrt{(1+x)}+\sqrt{(1+2x)}+\sqrt{x}}{\sqrt{(1+x)}-\sqrt{(1+2x)}+\sqrt{x}},$$

$4y^2(1+2x)^2 = (y-1)^4x(x+1).$

HINT. The methods of Example 6 are guides to a correct start.

4. Functions of two or more variables

Suppose that x, y, z are three distinct variables and that they are related in such a way that the value of z is fixed once the values of x and y have been fixed. Then we say that z is a function of the two variables x and y. For example, when

$$z = x^2+3xy-y^3, \tag{1}$$

z is a function of x and y: if we fix the values of x and y to be, say, 2 and -1, we thereby fix the value of z to be $4-6+1$, i.e. -1.

The expression on the right-hand side of (1) is a simple example of a POLYNOMIAL FUNCTION OF THE TWO VARIABLES x and y, or, briefly, 'A POLYNOMIAL IN x AND y'. The most general polynomial in x and y is a sum of a number of terms each of the form

$$a_{rs}x^ry^s,$$

where r, s are positive integers, or zero, and each a_{rs} is a constant.

Functions of three and more variables are similarly defined. We shall be concerned only with particular examples of such functions.

5. Notation for functions

5.1. For the sake of brevity we frequently use the notation $f(x)$ to denote a function of x: in using this notation $f(2)$ denotes the value of the function when x has the value 2, $f(\frac{3}{4})$ the value of the function when x has the value $\frac{3}{4}$, and so on. In order to state, at the beginning of a sum or problem, exactly what function we are considering, we shall use the identity sign i.e. $\equiv$. Thus, we write, for example,

$$f(x) \equiv 3x^2+7, \tag{1}$$

which is read '$f(x)$ is identically equal to $3x^2+7$', to mean that whatever value is assigned to the variable x, the symbol $f(x)$ will denote $3x^2+7$. Consequently, we then use such symbols as (i) $f(2)$ and (ii) $f(1-x)$ to denote the value of $3x^2+7$ when

(i) x is replaced by 2; that is,

$$3.4+7 = 19;$$

(ii) x is replaced by $1-x$; that is,

$$3(1-x)^2+7 = 3(1-2x+x^2)+7 = 10-6x+3x^2.$$

5.2. *Worked examples.*

(i) *When* $f(x) \equiv (3x-4)/(6x^3-7)$, *find the value of* $f(-2)$.

$$f(-2) = \frac{3(-2)-4}{6(-2)^3-7} = \frac{-6-4}{-48-7} = \frac{10}{55} = \frac{2}{11}.$$

(ii) *Prove that, if a, b, c are constants and if*

$$f(x) \equiv ax^2+bx+c,$$

then $f(x+1)-2f(x)+f(x-1) \equiv 2a$ (*that is, is equal to* $2a$ *whatever value is assigned to* x).

When $$f(x) \equiv ax^2+bx+c,$$

we have

$$\begin{aligned} &f(x+1)-2f(x)+f(x-1) \\ &\quad \equiv a(x+1)^2+b(x+1)+c-2(ax^2+bx+c) \\ &\qquad\qquad +a(x-1)^2+b(x-1)+c \\ &\quad \equiv a(x^2+2x+1)+b(x+1)+c-2(ax^2+bx+c) \\ &\qquad\qquad +a(x^2-2x+1)+b(x-1)+c \qquad (1) \\ &\quad \equiv 2a, \end{aligned}$$

the total coefficients of x^2 and of x in (1) being

$$a-2a+a, \quad \text{i.e. zero,}$$

and $$2a+b-2b-2a+b, \quad \text{i.e. zero.}$$

NOTE. We may use the identity sign, since each step of the argument holds whatever value is assigned to x. When we are considering *particular* values of x, we must not use the identity sign: for instance, we write

$$f(3)-2f(2)+f(1) = 9a+3b+c-2(4a+2b+c)+a+b+c$$
$$= 2a.$$

5.3. We may use letters other than f to denote 'function', just as we may use letters other than x to denote a 'variable'. For example, we may use† $F(x)$, $\phi(x)$, or $u(x)$ to denote a function of x; or again, $G(y)$, $h(y)$, or $\psi(y)$ to denote a function of y.

A function of the two variables x and y is commonly denoted by $f(x, y)$, $F(x, y)$, and so on.

5.4. *Worked example.*

PROBLEM.

When $f(x,y) \equiv x^2+y^2-2(x+y)$, *prove that*

$$f(2+x, 2+y)-f(x,y) \equiv 4(x+y).$$

SOLUTION.‡

$$f(2+x, 2+y) \equiv (2+x)^2+(2+y)^2-2(4+x+y)$$
$$\equiv 4+4x+x^2+4+4y+y^2-2(4+x+y)$$
$$\equiv 2(x+y)+x^2+y^2,$$
$$f(x,y) \equiv -2(x+y)+x^2+y^2.$$
$$\therefore \quad f(2+x, 2+y)-f(x,y) \equiv 4(x+y).$$

EXAMPLES I B

1. Find the values of

(i) $f(2)$, (ii) $f(-3)$, (iii) $f(\frac{1}{2})$

when $f(x) \equiv (3x^2+7x+2)/(3x-7)$.

2. Find the values of

(i) $f(3)$, (ii) $f(-1)$, (iii) $f(\frac{3}{2})$

when $f(x) \equiv (3x^2+7x+2)(4x-1)$.

3. (i) Find the value of the function

$$f(x,y) \equiv (x^2+y^2)-(x+y)$$

when $x = 2$ and $y = 3$.

† ϕ and ψ are the Greek letters 'phi' and 'psi'.

‡ The second line of working can be omitted when the student is sufficiently practised.

(ii) Prove that $f(1-x, 1-y)$, that is to say, the value of the function when x is replaced by $1-x$ and y is replaced by $1-y$, is identically equal to $f(x,y)$.

4. When $f(x) \equiv Ax+B$, where A and B are constants, prove that

$$f(x+1)-f(x) \equiv A,$$

and deduce that

$$f(x+2)-2f(x+1)+f(x) \equiv \{f(x+2)-f(x+1)\}-\{f(x+1)-f(x)\} \equiv 0.$$

5. Prove that, if $f(x) \equiv ax^2+bx+c$, where a, b, c are constants, and if $\phi(x) \equiv f(x+1)-f(x)$, then $\phi(x)$ is of the form $Ax+B$, where A and B are constants.

ANS. $A = 2a$, $B = a+b$. Notice that if $f(x)$ is any quadratic expression in x, then $f(x+1)-f(x)$ is linear in x: it is this fact that matters and not the values of A and B in terms of a and b.

6.* In the notation of Example 5 prove that $\phi(x+1)-\phi(x)$ is a constant, i.e. $f(x+2)-2f(x+1)+f(x)$ is a constant.

7.* By considering the result of Example 6 (a) as it stands, and (b) when x is replaced by $x+1$, and subtracting the two results, prove that

$$f(x+3)-3f(x+2)+3f(x+1)-f(x) \equiv 0,$$

when $f(x) \equiv ax^2+bx+c$.

8. When $f(x) \equiv ax^3+b$, where a and b are constants, prove that

(i) $f(x+2)-f(x)$ is divisible by 2,

(ii) $f(x+y)-f(x)$ is divisible by y.

9. When $f(x) \equiv ax^3+bx^2+cx+d$, where a, b, c, d are constants, prove that $f(x)-f(y)$ is divisible by $x-y$.

10.* Prove that, when $f(x)$ is a polynomial in x, $f(x)-f(y)$ is divisible by $x-y$.

11. When $f(x) \equiv ax+b$, prove that

$$f(x^2)-2f(xy)+f(y^2) \equiv a(x-y)^2.$$

SOLUTION.

$$f(x^2) \equiv ax^2+b,$$
$$-2f(xy) \equiv -2axy-2b,$$
$$f(y^2) \equiv ay^2+b.$$
$$\therefore\ f(x^2)-2f(xy)+f(y^2) \equiv a(x^2-2xy+y^2) \equiv a(x-y)^2.$$

12. When $f(x) \equiv ax+b$, prove that

$$f(x^3)-3f(x^2y)+3f(xy^2)-f(y^3) \equiv a(x-y)^3.$$

13. When $f(x) \equiv ax^2+bx+c$, prove that

$$\frac{f(x)-f(y)}{x-y}-\frac{f(y)-f(z)}{y-z} \equiv a(x-z).$$

CHAPTER II

POLYNOMIALS: GENERAL THEORY

1. Preliminary

Let $p_0, p_1, ..., p_n$, of which the first, p_0, is not zero, be given constants; let x be a variable. Then

$$p_0 x^n + p_1 x^{n-1} + p_2 x^{n-2} + ... + p_n \qquad (1)$$

is a polynomial of degree n in x. We shall denote this polynomial by $P(x)$.

We first examine the result of dividing $P(x)$ by $x-a$, where a is a constant. The first step of the usual long-division method, namely,

$$\begin{array}{l} x-a)\, p_0 x^n + p_1 x^{n-1} + ... + p_n\, (p_0 x^{n-1} + ... \\ \quad\quad\; \underline{p_0 x^n - a p_0 x^{n-1}} \\ \quad\quad\quad\quad (p_1 + a p_0) x^{n-1} + p_2 x^{n-2}, \end{array}$$

shows that the quotient when $x-a$ divides $P(x)$ is a polynomial of degree $n-1$, of which the first term is $p_0 x^{n-1}$.

We could, with care, complete the division, continuing the long-division process until the remainder contained no x. We should find that the final remainder was

$$p_0 a^n + p_1 a^{n-1} + p_2 a^{n-2} + ... + p_n. \qquad (2)$$

But the work would be long and many students would fail to follow it; so we find the remainder by other means. We must, however, note carefully that the final remainder would be a constant, involving no x; for we go on dividing and obtaining fresh terms in the quotient just so long as the remainder at any stage in the long-division process has an x in it; the process stops when no x is left in the remainder.

For example, when we divide

$$x^3 + 7x^2 - 3x + 10 \quad \text{by} \quad x - 1,$$

we set out the division sum as follows:

$$\begin{array}{l} x-1)\, x^3 + 7x^2 - 3x + 10\, (x^2 + 8x + 5 \\ \quad\quad\; \underline{x^3 - x^2} \\ \quad\quad\quad 8x^2 - 3x \\ \quad\quad\quad \underline{8x^2 - 8x} \\ \quad\quad\quad\quad\; 5x + 10 \\ \quad\quad\quad\quad\; \underline{5x - \;\, 5} \\ \quad\quad\quad\quad\quad\;\; \underline{\underline{15.}} \end{array}$$

We go on dividing until we are left with a remainder, 15, which contains no x: this is the final remainder and is a constant.

Moreover, since x^2+8x+5 is the quotient and 15 is the remainder,

$$x^3+7x^2-3x+10 \equiv (x-1)(x^2+8x+5)+15,$$

the identity sign being used because the two expressions are equal whatever be the value of x; each step of the long-division process is valid whatever x may be.

2. The Remainder Theorem

2.1. THEOREM 1. *The remainder when the polynomial*

$$P(x) \equiv p_0x^n+p_1x^{n-1}+...+p_n$$

is divided by $x-a$ is $P(a)$, i.e.

$$p_0a^n+p_1a^{n-1}+...+p_n.$$

This theorem is called the 'Remainder Theorem'.

PROOF. Let $Q(x)$, a polynomial of degree $n-1$ in x, be the quotient, and let R, a constant, be the remainder when $P(x)$ is divided by $x-a$. Then

$$P(x) \equiv (x-a)Q(x)+R, \tag{3}$$

the identity sign being used because the result is true for all values of x. Since (3) is true for all values of x, it is true when $x = a$. Put $x = a$ in (3); we obtain

$$P(a) = R.$$

Hence R, the remainder, is equal to $P(a)$.

2.2 *An alternative proof.* The remainder theorem can also be proved by actually carrying out the division. Let us divide

$$p_0x^3+p_1x^2+p_2x+p_3$$

by $x-a$. The working is

$$x-a)\,p_0x^3+p_1x^2+p_2x+p_3\,(p_0x^2+\overline{p_1+ap_0}\,x+\overline{p_2+ap_1+a^2p_0}$$

$$\underline{p_0x^3-ap_0x^2}$$

$$(p_1+ap_0)x^2+p_2x$$

$$\underline{(p_1+ap_0)x^2-(ap_1+a^2p_0)x}$$

$$(p_2+ap_1+a^2p_0)x+p_3$$

$$\underline{(p_2+ap_1+a^2p_0)x-(ap_2+a^2p_1+a^3p_0)}$$

$$\underline{\underline{p_3+ap_2+a^2p_1+a^3p_0.}}$$

The quotient is $p_0x^2+(p_1+ap_0)x+(p_2+ap_1+a^2p_0)$ and the remainder $p_3+ap_2+a^2p_1+a^3p_0$, which is the same as $p_0a^3+p_1a^2+p_2a+p_3$; hence, the remainder is the result of putting a instead of x in the dividend. This proves Theorem 1 when $n = 3$.

When $x-a$ divides $p_0x^n+p_1x^{n-1}+...+p_n$ the quotient runs

$$p_0x^{n-1}+(p_0a+p_1)x^{n-2}+(p_0a^2+p_1a+p_2)x^{n-3}+...;$$

and the remainder is $p_0a^n+p_1a^{n-1}+...+p_n$. It is a good exercise to work out the first few steps of the division, note how the terms of the quotient build themselves up in the regular pattern p_0x^n, $(p_0a+p_1)x^{n-1}$, $(p_0a^2+p_1a+p_2)x^{n-2}$,..., and so convince oneself that the first remainder to be free of x in the long-division process is $p_0a^n+p_1a^{n-1}+...+p_n$.

2.3. THEOREM 2. *If $P(a) = 0$, then the polynomial*

$$P(x) \equiv p_0x^n+p_1x^{n-1}+...+p_n$$

has $x-a$ as a factor; conversely, if $x-a$ is a factor of $P(x)$, then $P(a) = 0$.

PROOF. By Theorem 1, the remainder when $x-a$ divides $P(x)$ is $P(a)$. Therefore, if $P(a) = 0$, the remainder is zero and $x-a$ divides $P(x)$ exactly, i.e. $x-a$ is a factor of $P(x)$. Conversely, if $x-a$ is a factor of $P(x)$, the remainder must be zero; that is, $P(a)$ must be zero.

3. Applications of Theorems 1 and 2

We work some examples to show how various problems can be solved by using the Remainder Theorem.

PROBLEM 1. *Find the factors of*

$$2x^3-7x^2+7x-2.$$

Since the constant term is -2, the linear factors, if there are any, must have a constant term ± 1 or ± 2, for 1 and 2 are the only factors of 2. We therefore see whether any one of $x\pm 1$, $x\pm 2$ is a factor.

SOLUTION (i). Write

$$f(x) \equiv 2x^3-7x^2+7x-2.$$

Then $f(1) = 0$, $f(-1) = -2-7-7-2 \neq 0$.

Therefore (by Theorem 2) $x-1$ is a factor of $f(x)$, while $x+1$ is not a factor.

Divide $f(x)$ by the factor found, i.e. $x-1$. We get $2x^2-5x+2$, and

$$2x^2-5x+2 \equiv (2x-1)(x-2).$$

The required factors are, therefore, $x-1$, $2x-1$, $x-2$.

In this solution we divide out by one factor as soon as we find it and then factorize the quotient.

SOLUTION (ii).

$$\text{Write } f(x) \equiv 2x^3-7x^2+7x-2.$$

Then $\quad f(1) = 0, \qquad f(-1) = -2-7-7-2 \neq 0,$

$$f(2) = 16-28+14-2 = 0.$$

Therefore $x-1$ and $x-2$ are factors, and so their product, x^2-3x+2, is a factor. Divide $f(x)$ by x^2-3x+2. The result is $2x-1$. The factors are $x-1$, $x-2$, $2x-1$.

In this solution we use Theorem 2 to find as many factors as we can and then divide by the product of the factors so found.

PROBLEM 2. *Find the factors of*

$$3a^3-22a^2b+43ab^2-12b^3. \tag{i}$$

For the purpose of applying the result of Theorem 2 to this problem we think of (i) as a polynomial of degree 3 in a whose coefficients are 3, $-22b$, $43b^2$, and $-12b^3$. We write

$$f(a) \equiv 3a^3-22a^2b+43ab^2-12b^3$$

and, accordingly, we use the notations

$f(2)$ to mean $3.8-22.4b+43.2b^2-12b^3$,

$f(2b)$ to mean $3.8b^3-22.4b^2.b+43.2b.b^2-12b^3$, i.e. $10b^3$,

and so on.

SOLUTION.

$$\text{Write } f(a) \equiv 3a^3-22a^2b+43ab^2-12b^3.$$

Then,

$$f(b) = 3b^3-22b^3+43b^3-12b^3 \neq 0,$$

$$f(2b) = 10b^3 \neq 0,$$

$$f(3b) = 81b^3-22.9b^3+43.3b^3-12b^3$$

$$= (81-198+129-12)b^3 = 0.$$

$\therefore$ $a-3b$ is a factor and the quotient, on dividing out by $a-3b$, is† $3a^2-13ab+4b^2$. The factors of this are $(3a-b)(a-4b)$.

Hence the factors are $a-3b$, $a-4b$, and $3a-b$.

† The reader must work out the division sum for himself.

PROBLEM 3. *The polynomial*

$$x^8+2x^7+3x^2+ax+b \qquad \text{(i)}$$

is divisible without remainder by x^2+x-2. *Find the values of* a *and* b.

SOLUTION.

$$x^2+x-2 \equiv (x+2)(x-1),$$

and therefore $x+2$ and $x-1$ are factors of (i). Hence, when we put $x=-2$ or $x=1$, the value of (i) is zero.

$$\therefore\quad (-2)^8+2(-2)^7+3(-2)^2-2a+b = 0,$$

and
$$1+2+3+a+b = 0.$$

That is,
$$-2a+b = -12$$

and
$$a+b = -6.$$

Solving these two equations, we get $a = 2$, $b = -8$.

PROBLEM 4. *Prove, by means of the remainder theorem, that* $2x-1$ *is a factor of* $2x^6-3x^5+x^4-2x^2+3x-1$.

SOLUTION.

Write $f(x) \equiv x^6-\frac{3}{2}x^5+\frac{1}{2}x^4-x^2+\frac{3}{2}x-\frac{1}{2}$.

Then
$$f(\tfrac{1}{2}) = \tfrac{1}{64}-\tfrac{3}{64}+\tfrac{1}{32}-\tfrac{1}{4}+\tfrac{3}{4}-\tfrac{1}{2} = 0,$$

and so $x-\frac{1}{2}$ is a factor of $f(x)$.

Hence $2x-1$ is a factor of $2\times f(x)$, i.e. is a factor of

$$2x^6-3x^5+x^4-2x^2+3x-1.$$

EXAMPLES II A

1. Find the remainder when
   (i) $3x^3+7x+6$ is divided by $x+1$,
   (ii) $5x^4+14x^2+5$ is divided by $x-2$,
   (iii) x^3+x^2+x+1 is divided by $x+3$,
   (iv) x^3+x^2-x+1 is divided by $x+2$.

2. Find the value of a
   (i) when $3x^3+ax^2-7x+6$ is divisible without remainder by $x-1$;
   (ii) when $5x^4+ax^3+6x-4$ is divisible without remainder by $x+2$.

3. Find the values of a and b when
   (i) $3x^3-4x^2+ax+b$ is divisible without remainder by x^2-1;
   (ii) $6x^3+ax^2+bx-2$ is divisible without remainder by $2x-1$ and by $x+1$.

4. Show that when a polynomial $f(x)$, of degree n in x, is divided by $(x-1)(x-2)$, the remainder is of the form $ax+b$, where a and b are constants; and deduce that

$$f(x) \equiv (x-1)(x-2)Q(x)+ax+b,$$

where $Q(x)$ is a polynomial of degree $n-2$ in x.

5. The remainder when x^2-3x+2 divides $px^4+qx^3-18x^2+15x-5$ is $4x-7$. Show that $p=1$ and $q=4$.

HINT.

$$px^4+qx^3-\ldots \equiv (x-1)(x-2)(px^2+\ldots)+4x-7, \qquad \text{(i)}$$

where $px^2+\ldots$ denotes the quotient when the left-hand side is divided by x^2-3x+2, i.e. $(x-1)(x-2)$.

In (i) put $x=1$, $x=2$ and solve the resulting equations for p and q.

6. (i) The remainder when $(x+2)(x+3)$ divides x^5+ax^2+b is $x+1$. Show that $a=42$ and $b=-137$.

(ii) The remainder when x^2-1 divides x^6+ax^3+b is $2x+3$. Show that $a=b=2$.

7.* Given that $x+2$ is a factor of

$$x^4-bx^3-11x^2+4(b+1)x+a$$

and that the expression itself is the square of a quadratic in x, find the values of a and b.

8. Find the factors of

(i) $2x^3-3x^2-3x+2$, (ii) $6x^3-x^2-19x-6$.

9. Find the factors of

(i) $x^3-6x^2+11x-6$, (ii) x^3-7x+6, (iii) $x^4+x^3-7x^2-x+6$.

10. Find the factors of

(i) $x^3-6ax^2+11a^2x-6a^3$, (ii) $x^3-7xy^2+6y^3$.

11. Find the factors of

(i) $2a^3-3a^2b-3ab^2+2b^3$, (ii) $6b^3-b^2c-19bc^2-6c^3$.

12. Prove that

(i) $3x-1$ is a factor of $3x^3-22x^2+43x-12$;

(ii) $2a+b$ is a factor of $10a^3+17a^2b+20ab^2+7b^3$.

4. A polynomial as the product of factors

4.1. *An important detail.* Let a, b be two numbers, neither of them being zero. Then their product ab is not zero. Hence, when the product of two numbers is known to be zero, the two numbers cannot both be different from zero.

For example, if $cd = 0$ and it is known that $d \neq 0$, then it follows that $c = 0$.

Again, if $ad = cd$ and it is known that $d \neq 0$, then $a = c$; for when $ad = cd$ we have

$$(a-c)d = 0,$$

and so, since $d \neq 0$, we have $a-c = 0$, or $a = c$. Thus, if

$$ad = cd \quad (d \neq 0),$$

then

$$a = c.$$

In words, WE CAN DIVIDE BOTH SIDES OF AN EQUATION BY ANY FACTOR WHICH IS NOT EQUAL TO ZERO.

On the other hand, it does not follow from the known fact $0 \times 7 = 0 \times 6$, that 7 is equal to 6. It is quite improbable that anyone *would* divide by 0, once the zero was clearly set down as such on the paper. The danger is that when working with symbols like a, p_0, x, which may denote any numbers, we are liable to divide by them and forget to note that the result so obtained may not be true when the symbol stands for the particular value zero.

4.2. *Fundamental theorems.* Theorems 3–5, which follow, are are of great importance in the theory of algebra. The proofs of these theorems may seem difficult. The reader is advised (i) to follow the line of argument that leads up to Theorem 5 and its corollaries as well as he can (many readers will have no particular difficulty), (ii) to master the facts contained in the enunciations of Theorem 5 and its corollaries, and then to study the worked examples in § 5.

THEOREM 3. *If the polynomial*

$$P(x) \equiv p_0 x^n + p_1 x^{n-1} + \ldots + p_n \quad (p_0 \neq 0)$$

is equal to zero when x has any one of the n distinct values a_1, $a_2, \ldots, a_n$, then

$$p_0 x^n + p_1 x^{n-1} + \ldots + p_n \equiv p_0(x-a_1)(x-a_2)\ldots(x-a_n). \quad (1)$$

The identity sign $\equiv$ between the two sides of (1) means that the two sides are equal for all values of the variable x.

PROOF. By hypothesis,

$$P(a_1) = 0, \qquad P(a_2) = 0, \qquad ..., \qquad P(a_n) = 0.$$

Since $P(a_1) = 0$, $x-a_1$ is a factor of $P(x)$. Moreover, as we noted in § 1.1, the quotient when $x-a_1$ divides $P(x)$ is a polynomial of degree $n-1$ whose first term is $p_0 x^{n-1}$. Let us denote this quotient by $Q_{n-1}(x)$. Then we have, so far, proved that

$$P(x) \equiv (x-a_1)Q_{n-1}(x), \tag{2}$$

where $Q_{n-1}(x) \equiv p_0 x^{n-1}+\ldots$.

Since $P(a_2) = 0$, we have, on putting $x = a_2$ in (2),

$$0 = P(a_2) = (a_2-a_1)Q_{n-1}(a_2). \tag{3}$$

Now $a_2 \neq a_1$ (by hypothesis) and therefore, by the argument of § 4.1,

$$Q_{n-1}(a_2) = 0.$$

Hence $x-a_2$ is a factor of $Q_{n-1}(x)$.

The quotient when $x-a_2$ divides $Q_{n-1}(x)$ will be $p_0 x^{n-2}+\ldots$. Denote this quotient by $Q_{n-2}(x)$. Then

$$Q_{n-1}(x) \equiv (x-a_2)Q_{n-2}(x),$$

and so

$$P(x) \equiv (x-a_1)Q_{n-1}(x)$$
$$\equiv (x-a_1)(x-a_2)Q_{n-2}(x), \tag{4}$$

where $Q_{n-2}(x) \equiv p_0 x^{n-2}+\ldots$.

We may continue the process of taking out factors $x-a_1$, $x-a_2$, $x-a_3$,.... When we have taken out one factor $x-a_1$, as in (2), the quotient is $Q_{n-1}(x)$ or $p_0 x^{n-1}+\ldots$; when we have taken out two factors $x-a_1$ and $x-a_2$, as in (4), the quotient is $Q_{n-2}(x)$ or $p_0 x^{n-2}+\ldots$. Thus when we have taken out n factors $x-a_1$, $x-a_2$,..., $x-a_n$, the quotient will be $Q_0(x)$ or $p_0 x^0$, which is simply p_0 (since $x^0 \equiv 1$). Hence

$$P(x) \equiv p_0(x-a_1)(x-a_2)\ldots(x-a_n).$$

4.3. THEOREM 4. *A polynomial of degree n in x cannot be equal to zero for more than n distinct values of x.*

PROOF. Let the polynomial of degree n be

$$P(x) \equiv p_0 x^n+p_1 x^{n-1}+\ldots+p_n \quad (p_0 \neq 0).$$

The condition $p_0 \neq 0$ is necessary in order to ensure that $P(x)$ is, in fact, of degree n and not of some less degree.

Let $P(x)$ be zero when x has the n distinct values $a_1, a_2,..., a_n$. Then, by Theorem 3,

$$P(x) \equiv p_0(x-a_1)(x-a_2)...(x-a_n). \tag{5}$$

Moreover, (5) cannot be zero when x has any value distinct from $a_1, a_2,..., a_n$; for no factor of (5) is zero for such a value of x and, by hypothesis, p_0 is not zero.

4.4. THEOREM 5. *Given that the expression*

$$p_0 x^n + p_1 x^{n-1} + ... + p_n$$

is equal to zero for more than n distinct values of x, it follows that

$$p_0 = p_1 = ... = p_n = 0,$$

and that
$$p_0 x^n + p_1 x^{n-1} + ... + p_n \equiv 0.$$

PROOF.

EITHER $p_0, p_1,..., p_n$ are all zero,

OR there is a first one in the set that is not zero.

We show that the second alternative would lead to a contradiction and is therefore impossible.

Suppose p_k, where $k < n$, is the first of $p_0, p_1,...$ to be different from zero. Then the expression reduces to

$$p_k x^{n-k} + p_{k+1} x^{n-k-1} + ... + p_n \quad (p_k \neq 0).$$

This is a polynomial of degree $n-k$ in x. Therefore, by Theorem 4, it cannot be equal to zero for more than $n-k$ values of x. But we are given that the expression is equal to zero for more than $n-k$ values of x and therefore our supposition has led to a contradiction. Hence the supposition is impossible.

Now suppose that p_n is the first of $p_0, p_1,...$ to be different from zero. Then the expression $p_0 x^n + p_1 x^{n-1} + ... + p_n$ reduces simply to p_n, which we are supposing to be different from zero. Thus the expression is never equal to zero, whatever the value of x. This contradicts the original hypothesis that the expression is zero for more than n distinct values of x. Hence the supposition is impossible.

Accordingly, only the first alternative, i.e.

$$p_0 = p_1 = ... = p_n = 0,$$

is possible. Further, when $p_0 = p_1 = ... = p_n = 0$, the expression
$$p_0x^n+p_1x^{n-1}+...+p_n$$
is equal to zero for all values of x. This proves the theorem.

COROLLARY 1. *Given that*
$$p_0x^n+p_1x^{n-1}+...+p_n \equiv 0,$$
it follows that $$p_0 = p_1 = ... = p_n = 0.$$

PROOF. If the expression $p_0x^n+p_1x^{n-1}+...$ is identically zero, it is zero for more than n values of x.

COROLLARY 2. *If, for all values of x or for more than n values of x,*
$$p_0x^n+p_1x^{n-1}+...+p_n = q_0x^n+q_1x^{n-1}+...+q_n, \tag{1}$$
then $$p_0 = q_0,\ p_1 = q_1,...,\ p_n = q_n.$$

PROOF. If (1) is true for more than n values of x, then so is
$$(p_0-q_0)x^n+(p_1-q_1)x^{n-1}+...+(p_n-q_n) = 0,$$
and hence $p_0-q_0 = 0,\ p_1-q_1 = 0,...,\ p_n-q_n = 0.$

4.5. *Equating coefficients.* The process of deducing from the identity
$$p_0x^n+p_1x^{n-1}+...+p_n \equiv q_0x^n+q_1x^{n-1}+...+q_n$$
the facts that $p_0 = q_0,\ p_1 = q_1,...,\ p_n = q_n$ is commonly called 'EQUATING COEFFICIENTS'. The process is widely used.

5. Applications of Theorems 4 and 5

PROBLEM 1. *Show that constants a, b, c, d can be found such that*
$$n^3 \equiv a(n+1)(n+2)(n+3)+b(n+1)(n+2)+c(n+1)+d, \tag{1}$$
and find their values.

SOLUTION.
$$a(n+1)(n+2)(n+3)+b(n+1)(n+2)+c(n+1)+d$$
$$\equiv a(n^3+6n^2+11n+6)+b(n^2+3n+2)+c(n+1)+d$$
$$\equiv an^3+n^2(6a+b)+n(11a+3b+c)+(6a+2b+c+d).$$
This is identically equal to n^3 if
$$a = 1,\quad 6a+b = 0,\quad 11a+3b+c = 0,\quad 6a+2b+c+d = 0;$$
i.e. if
$$a = 1,\ b = -6,\ c = -11+18 = 7,\ d = -6+12-7 = -1.$$

$\therefore$ there are such constants a, b, c, d and they are

$$a = 1, \quad b = -6, \quad c = 7, \quad d = -1;$$

i.e. $n^3 \equiv (n+1)(n+2)(n+3)-6(n+1)(n+2)+7(n+1)-1$.

We may check the accuracy of this result by inserting particular values of n.

Check by $n = 0$; $\quad 0 = 6-12+7-1$.

Check by $n = -1$; $-1 = 0-0+0-1$.

PROBLEM 2. *Show that no constants a and b can be found such that*

$$n^2 \equiv a(n+1)(n+2)+b(n+2)(n+3). \qquad (2)$$

SOLUTION.

$$\begin{aligned} a(n+1)(n+2)+b(n+2)(n+3) & \\ &\equiv a(n^2+3n+2)+b(n^2+5n+6) \\ &\equiv n^2(a+b)+n(3a+5b)+(2a+6b). \end{aligned}$$

This will be identically equal to n^2 only if

$$a+b = 1, \quad 3a+5b = 0, \quad 2a+6b = 0.$$

The only solution of the last two equations is $a = 0$, $b = 0$, and these values do not satisfy the first equation. Hence no values can be found to satisfy the required conditions.

PROBLEM 3. *Prove that, when $a \neq b \neq c$,*

$$\frac{a^2(x-b)(x-c)}{(a-b)(a-c)}+\frac{b^2(x-c)(x-a)}{(b-c)(b-a)}+\frac{c^2(x-a)(x-b)}{(c-a)(c-b)} \equiv x^2. \qquad (3)$$

SOLUTION. When $x = a$, the left-hand side and the right-hand side of (3) both equal a^2: hence the left-hand and right-hand sides of (3) are equal when $x = a$. Similarly, they are equal when $x = b$ and when $x = c$.

The two quadratic expressions are equal for 3 distinct values of the variable x. They are therefore identically equal (Theorem 5, Corollary 2).

PROBLEM 4. *Prove that, if x^3+1 is a factor of*

$$x^5+ax^4+bx^3+cx^2+dx+e,$$

then $a = d$, $b = e$, and $c = 1$.

SOLUTION. If x^3+1 is one factor, there must be another factor of the form x^2+px+e, and then

$$x^5+ax^4+bx^3+cx^2+dx+e \equiv (x^3+1)(x^2+px+e)$$
$$\equiv x^5+px^4+ex^3+x^2+px+e.$$

On equating coefficients,

$$a = p, \qquad b = e, \qquad c = 1, \qquad d = p,$$

and hence $a = d$, $b = e$, and $c = 1$.

EXAMPLES II B

1. Find the values of a, b, c, d which are such that

(i) $x^2 \equiv a(x+2)^2+b(x+2)+c$,

(ii) $x^3 \equiv a(x+1)^3+b(x+1)^2+c(x+1)+d$,

(iii) $3x^3+2x^2+x-1 \equiv a(x-1)^3+b(x-1)^2+c(x-1)+d$.

2.* Show that any cubic polynomial in x can be written as a cubic polynomial in $x-1$.

3.** Show that any polynomial in x can be written as a polynomial of equal degree in $x-h$, where h is a given constant.

4. Find the values of a, b, c, d which are such that

(i) $n^2 \equiv a(n-1)n+bn$;

(ii) $n^3 \equiv a(n-1)n(n+1)+b(n-1)+c$;

(iii) $n^3+3n^2+5n-7 \equiv a(n-1)n(n+1)+b(n-1)n+c(n-1)+d$.

5. Find the values of a, b, c, d which are such that

(i) $x^3 \equiv a(x+2)^3+b(x+1)^2+cx+d$;

(ii) $x^2 \equiv a(x+1)^2+bx+c$;

(iii) $x^3+2x^2-7 \equiv a(x+1)^3+bx^2+c(x-1)+d$.

6. Prove that there are no values for a, b, c which are such that

$$x^3 \equiv a(x-1)x(x+1)+bx(x+1)(x+2)+c(x+1)(x+2)(x+3),$$

but that

$$x^3 \equiv \tfrac{7}{6}(x-1)x(x+1)-\tfrac{1}{2}x(x+1)(x+2)+\tfrac{1}{6}(x+1)(x+2)(x+3)-1.$$

7. Prove that, when $a \neq b \neq c$,

$$\frac{(x-b)(x-c)}{(a-b)(a-c)}+\frac{(x-c)(x-a)}{(b-c)(b-a)}+\frac{(x-a)(x-b)}{(c-a)(c-b)} \equiv 1,$$

$$\frac{a(x-b)(x-c)}{(a-b)(a-c)}+\frac{b(x-c)(x-a)}{(b-c)(b-a)}+\frac{c(x-a)(x-b)}{(c-a)(c-b)} \equiv x.$$

8. Prove that there cannot be two different quadratic expressions in x which have the values A, B, C when x has the values a, b, c; and show that the one quadratic that has these values is

$$A\frac{(x-b)(x-c)}{(a-b)(a-c)}+B\frac{(x-c)(x-a)}{(b-c)(b-a)}+C\frac{(x-a)(x-b)}{(c-a)(c-b)}.$$

HINT. If px^2+qx+r and $p'x^2+q'x+r'$ both have the required values, then $(p-p')x^2+(q-q')x+(r-r')$ is zero when x has the three distinct values a, b, c.

9.* Prove that there is one and only one cubic in x that has the values A, B, C, D when x has the values a, b, c, d; and write down this cubic by analogy with the result of Example 8.

10. Prove that, if ax^3+bx^2+cx+d contains $(x-1)^2$ as a factor, then $b = d-2a$ and $c = a-2d$.

11. Prove that, if ax^3+bx^2+bx+d contains $(x-1)^2$ as a factor, then $a = d = -b$.

12. Prove that, if x^4-px+q contains a factor $(x-c)^2$, then

$$27p^4 = 256q^3 \quad \text{and} \quad 3c^4 = q.$$

13.* Prove that, if $27p^4 = 256q^3$, then x^4-px+q contains a factor of the form $(x-c)^2$.

HINT. Put $q = 3c^4$.

6. Polynomials in two or more variables

6.1. THEOREM 6. *If $a,\ldots, c, a',\ldots, c'$ are constants and*

$$ax^2+2hxy+by^2+2gx+2fy+c$$
$$\equiv a'x^2+2h'xy+b'y^2+2g'x+2f'y+c',$$

then the corresponding coefficients are equal; that is,

$$a = a', \qquad h = h', \qquad \ldots, \qquad c = c'.$$

PROOF. Each expression may be considered as a quadratic in x whose coefficients are polynomials in y; for example, the first expression is

$$ax^2+2x(hy+g)+(by^2+2fy+c).$$

Since the two expressions are equal for all values of x and y, we have (by Theorem 5, Corollary 2)

$$a = a', \quad hy+g \equiv h'y+g', \quad by^2+2fy+c \equiv b'y^2+2f'y+c'.$$

But since $hy+g \equiv h'y+g'$ we have (again by Theorem 5, Corollary 2) $h = h'$ and $g = g'$. Similarly, since

$$by^2+2fy+c \equiv b'y^2+2f'y+c',$$

we have $b = b'$, $f = f'$, $c = c'$.

Hence the corresponding coefficients are equal.

COROLLARY 1. *The theorem remains true when the polynomials are of any degrees in x and y.*

COROLLARY 2. *The theorem remains true when the polynomials are of any degrees in any number of variables, x, y, z,...; that is,* THE PROCEDURE OF EQUATING COEFFICIENTS CAN BE APPLIED TO ANY TWO POLYNOMIALS THAT ARE EQUAL IDENTICALLY.

COROLLARY 3. *Two polynomials of different degrees cannot be identically equal.*

7.* Identities; a point in some methods of proof

In order to prove that two given polynomials, say $f(x,y)$ and $F(x,y)$, are identical, it is often convenient to suppose in the course of the proof that some third polynomial, say $P(x,y)$, is not equal to zero. We shall show that the two polynomials $f(x,y)$ and $F(x,y)$ are equal for ALL values of x and y if they are equal whenever $P(x,y)$ is not zero. We shall give the argument in a simple case, but the reasoning is quite general and the more advanced reader will find it worth while to set out the argument in its general form.

Suppose we know that

$$ax^2+2hxy+by^2+2gx+2fy+c \quad \text{(A)}$$
$$= a'x^2+2h'xy+b'y^2+2g'x+2f'y+c' \quad \text{(A')}$$

whenever $lx+my+n \neq 0$. Then, for any given value of y, (A) = (A') for all values of x except, possibly, $-(my+n)/l$. It follows that (Theorem 5, Corollary 2)

$$a = a', \quad hy+g \equiv h'y+g', \quad by^2+2fy+c \equiv b'y^2+2f'y+c',$$

and so $a = a'$, $h = h'$,..., $c = c'$.

From the equality of the corresponding coefficients it follows that (A) = (A') for ALL values of x and y and not merely for those values which make $lx+my+n$ different from zero.

The general theorem may be proved in the same way. It runs:

THEOREM 7. *If two polynomials $f(x,y,z,...)$ and $F(x,y,z,...)$ are equal for all values of the variables which satisfy a given set of inequalities*

$$P_1(x,y,z,...) \neq 0, \quad ..., \quad P_k(x,y,z,...) \neq 0,$$

where each P denotes a polynomial, then $f(x,y,z,...)$ and $F(x,y,z,...)$ are identically equal.

The theorem is not necessary in elementary work, though it is often applicable in more advanced work, especially in establishing results concerning determinants. It is sometimes known by the rather high-sounding title 'THE PRINCIPLE OF THE IRRELEVANCE OF ALGEBRAIC INEQUALITIES'.

The following is an example of its application. From Examples II B, 7 (i) we see, on multiplying throughout by $(b-c)(c-a)(a-b)$, that whenever $a \neq b \neq c$,

$$(b-c)(x-b)(x-c)+(c-a)(x-c)(x-a)+(a-b)(x-a)(x-b)$$
$$= -(b-c)(c-a)(a-b). \quad (1)$$

Regard (1) as the equivalence of two polynomials in the four variables x, a, b, c. The equivalence has, so far, been proved subject to the inequalities $b-c \neq 0$, $c-a \neq 0$, $a-b \neq 0$. By Theorem 7 the equivalence is true both when these inequalities are satisfied and when they are not.

EXAMPLES II C

1. Given that $ax^3+bx^2y+cxy^2+dy^3$ is a numerical multiple of $(x-y)^3$, express b, c, d in terms of a.

2. Given that

$$ax^2+2hxy+by^2+2gx+2fy+c \equiv (lx+my+n)(l'x+m'y+n'),$$

prove that $4(bc-f^2) = -(mn'-m'n)^2$.

3. Prove that there are two values of f for which

$$2x^2-5xy-3y^2-x+2fy-3$$

is the product of two linear factors. State the two values of f.

HINT. $2x^2-5xy-3y^2 \equiv (2x+y)(x-3y)$, so that the two linear factors must be $(2x+y+p)(x-3y-3/p)$.

4. Find the values of a, b, c in order that $ax^2+2bxy+cy^2+2x+4y+1$ may be a perfect square $(lx+my+1)^2$.

5. Prove that there are two values of h for which

$$2x^2-3y^2+4z^2+yz+6zx-hxy$$

is the product of linear factors, and find these linear factors.

CHAPTER III

POLYNOMIALS: EQUATIONS

1. The roots of an equation

1.1. Let

$$f(x) \equiv p_0 x^n + p_1 x^{n-1} + \ldots + p_n \quad (p_n \neq 0)$$

be a given polynomial of degree n in x. Then the values of x for which

$$p_0 x^n + p_1 x^{n-1} + \ldots + p_n = 0 \tag{1}$$

are called the ZEROS of $f(x)$, or the ROOTS of the equation (1). We refer to (1) as 'an equation of the nth degree in x' or 'an equation of degree n in x'.

By Theorem 2, if a is a root of (1), $f(x)$ contains $x-a$ as a factor. If $x-a$ is a factor, but $(x-a)^2$ is not a factor of $f(x)$, we say that a is a SIMPLE ROOT or a non-repeated root. If $(x-a)^2$ is a factor, but $(x-a)^3$ is not, we say that a is a DOUBLE ROOT; and so on for a triple root or, generally, an r-ple root, that is, a root a such that $(x-a)^r$ is a factor of $f(x)$ while $(x-a)^{r+1}$ is not.

Suppose that a is a double root. Then we may write

$$f(x) \equiv p_0(x-a)^2(x^{n-2} + q_1 x^{n-3} + \ldots), \tag{2}$$

where $p_0(x^{n-2} + q_1 x^{n-3} + \ldots)$ is the quotient when $(x-a)^2$ divides $f(x)$. If now $x-b$, $x-c$,..., $x-k$ are $n-2$ distinct factors of $x^{n-2} + q_1 x^{n-3} + \ldots$, we may write (by Theorem 3)

$$f(x) \equiv p_0(x-a)^2(x-b)(x-c)\ldots(x-k).$$

In this case there are only $n-1$ distinct values of x which give $f(x)$, a polynomial of degree n, the value zero. In order to have n roots for an equation of degree n we say, in such a case, that the roots are a (twice), b, c,..., k.

Thus, since $x^4 - 2x^2 + 1 \equiv (x^2-1)^2 \equiv (x-1)^2(x+1)^2$, we say that the roots of the equation

$$x^4 - 2x^2 + 1 = 0$$

are 1, 1, -1, -1.

Generally, if

$$p_0x^n+p_1x^{n-1}+...+p_n \equiv p_0(x-a)(x-b)...(x-k), \qquad (3)$$

where $a, b,..., k$ are n numbers, equal or unequal, we say that the roots of the equation

$$p_0x^n+p_1x^{n-1}+...+p_n = 0$$

are $a, b,..., k$.

1.2. It can be proved, though not by any known elementary and easy method,† that every polynomial $f(x)$ of degree n can be expressed as the product of a constant and n linear factors of the type $x-a, x-b,..., x-k$, the factors being not necessarily distinct and the numbers $a, b,..., k$ being either real numbers or complex‡ numbers. The numbers $a, b,..., k$ are the n roots of the equation $f(x) = 0$.

2. Relations between roots and coefficients; quadratic and cubic equations

2.1. *Quadratic equations.* Consider first a quadratic equation

$$ax^2+bx+c = 0 \quad (a \neq 0),$$

whose roots are α and β. Then

$$ax^2+bx+c \equiv a(x-\alpha)(x-\beta). \qquad (1)$$

But $$(x-\alpha)(x-\beta) \equiv x^2-(\alpha+\beta)x+\alpha\beta$$

and so $$ax^2+bx+c \equiv ax^2-a(\alpha+\beta)x+a\alpha\beta.$$

On equating coefficients (Theorem 5, Corollary 2),

$$-a(\alpha+\beta) = b, \qquad a\alpha\beta = c,$$

or $$\alpha+\beta = -\frac{b}{a}, \qquad \alpha\beta = \frac{c}{a}.$$

We sum up these results in Theorem 8.

† The crux of the difficulty is to show that every equation of the form $p_0x^n+p_1x^{n-1}+...+p_n = 0$ $(p_0 \neq 0)$ has at least one root. Once this is established, it is not difficult to show that the equation has n roots and only n roots.

‡ The theorem is not true if we confine ourselves to real numbers, such as $1, -\frac{1}{2}, \sqrt{2}, \pi$. For example, there is no positive or negative number whose square is -1, so that the equation $x^2+1 = 0$ has no root among the positive or negative numbers. We introduce complex numbers in § 3.

THEOREM 8. *If the roots of the quadratic equation*

$$ax^2+bx+c = 0$$

are α *and* β, *then*

$$\alpha+\beta = -\frac{b}{a}, \qquad \alpha\beta = \frac{c}{a}. \tag{2}$$

Conversely, if α *and* β *are given to be the roots of a quadratic equation, then that equation is*† $x^2+px+q = 0$, *where*

$$p = -(\alpha+\beta), \qquad q = \alpha\beta.$$

We shall work one or two examples on the application of Theorem 8. Before doing so we call attention to the fact that in many problems of mathematics the two roots of a quadratic equation are so closely united that the problem *must* be worked by handling both roots at once, that is, by using Theorem 8, and *must not* be worked by the ugly and long-winded method which consists of solving the quadratic equation and substituting the values of the roots. Especially is this the case when quadratic equations occur in the problems of analytical geometry.

In elementary work one solves a quadratic equation; in all work that is the least advanced in character one should avoid the actual solution of the quadratic equation whenever possible: instead of solving the equation, one should 'suppose the roots to be α and β' and then make use of Theorem 8.

2.2. *Cubic equations.* Consider next a cubic equation

$$ax^3+bx^2+cx+d = 0$$

whose roots are α, β, γ. Then

$$ax^3+bx^2+cx+d \equiv a(x-\alpha)(x-\beta)(x-\gamma).$$

Now‡

$$(x-\alpha)(x-\beta)(x-\gamma)$$
$$\equiv (x-\alpha)\{x^2-(\beta+\gamma)x+\beta\gamma\} \tag{3}$$
$$\equiv x^3-x^2(\alpha+\beta+\gamma)+x(\beta\gamma+\gamma\alpha+\alpha\beta)-\alpha\beta\gamma. \tag{4}$$

† The equation $x^2+px+q = 0$ has the same roots as the equation

$$ax^2+apx+aq = 0;$$

for a 'root' is a value of x that makes x^2+px+q equal to zero, and such a value also makes $a(x^2+px+q)$ equal to zero whatever value is assigned to a.

‡ In working out (3) to obtain (4), pick out the one term x^3, then pick out all terms in x^2, then all terms in x, and, finally, the term independent of x. Note that we write $\beta\gamma+\gamma\alpha+\alpha\beta$ and NOT $\alpha\beta+\alpha\gamma+\beta\gamma$, or any other non-symmetrical arrangement.

Thus

$$ax^3+bx^2+cx+d \equiv ax^3-a(\alpha+\beta+\gamma)x^2+a(\beta\gamma+\gamma\alpha+\alpha\beta)x-a\alpha\beta\gamma$$

and, on equating coefficients (Theorem 5, Corollary 2),

$$\alpha+\beta+\gamma = -\frac{b}{a}, \qquad \beta\gamma+\gamma\alpha+\alpha\beta = \frac{c}{a}, \qquad \alpha\beta\gamma = -\frac{d}{a}.$$

These results give

THEOREM 9. *If the roots of the cubic equation*

$$ax^3+bx^2+cx+d = 0$$

are α, β, γ, *then*

$$\alpha+\beta+\gamma = -\frac{b}{a}, \qquad \beta\gamma+\gamma\alpha+\alpha\beta = \frac{c}{a}, \qquad \alpha\beta\gamma = -\frac{d}{a}. \tag{5}$$

Conversely, if α, β, γ *are given to be the roots of a cubic equation, then that equation is*

$$x^3+px^2+qx+r = 0,$$

where

$$p = -(\alpha+\beta+\gamma), \qquad q = \beta\gamma+\gamma\alpha+\alpha\beta, \qquad r = -\alpha\beta\gamma. \tag{6}$$

Notice the plus and minus signs in (5) and (6).

2.3. *Worked examples.*

PROBLEM 1. *Given that* α *and* β *are the roots of the quadratic equation* $ax^2+bx+c = 0$, *express* $\alpha^2+\beta^2$ *and* $\alpha^3+\beta^3$ *in terms of* a, b, c.

Essentially, the problem is that of expressing $\alpha^2+\beta^2$ and $\alpha^3+\beta^3$ in terms of $(\alpha+\beta)$ and $\alpha\beta$, since we know the values of the latter in terms of a, b, c.

SOLUTION. Since α and β are the roots of the equation $ax^2+bx+c = 0$,

$$\alpha+\beta = -\frac{b}{a}, \qquad \alpha\beta = \frac{c}{a}.$$

Now $$\alpha^2+\beta^2 = (\alpha+\beta)^2-2\alpha\beta;$$

$$\therefore \quad \alpha^2+\beta^2 = \frac{b^2}{a^2}-2\frac{c}{a} = \frac{b^2-2ac}{a^2}.$$

Also,
$$\alpha^3+\beta^3 = (\alpha+\beta)(\alpha^2-\alpha\beta+\beta^2)$$
$$= -\frac{b}{a}\{(\alpha+\beta)^2-3\alpha\beta\}$$
$$= -\frac{b}{a}\left(\frac{b^2}{a^2}-\frac{3c}{a}\right)$$
$$= -\frac{b(b^2-3ac)}{a^3}.$$

Problem 2. *One root of the equation $ax^2+bx+c = 0$ is twice the other root. Prove that $2b^2 = 9ac$.*

Solution. Let the roots be α and 2α. Then
$$3\alpha = -\frac{b}{a}, \qquad 2\alpha^2 = \frac{c}{a}.$$
$$\therefore \frac{b^2}{a^2} = 9\alpha^2 = \frac{9c}{2a}.$$
$$\therefore 2b^2 = 9ac.$$

Problem 3. *Given that α and β are the roots of the equation $x^2+5x+10=0$, construct the equation whose roots are $2\alpha-3\beta$ and $2\beta-3\alpha$.*

Solution. By hypothesis, $\alpha+\beta = -5$ and $\alpha\beta = 10$. Hence
$$(2\alpha-3\beta)+(2\beta-3\alpha) = -(\alpha+\beta) = 5, \tag{1}$$
and
$$(2\alpha-3\beta)(2\beta-3\alpha) = -6(\alpha^2+\beta^2)+13\alpha\beta$$
$$= -6\{(\alpha+\beta)^2-2\alpha\beta\}+13\alpha\beta$$
$$= -6(\alpha+\beta)^2+25\alpha\beta$$
$$= -6.25+250 = 100. \tag{2}$$
By (1) and (2), the required equation is (converse of Theorem 8)
$$x^2-5x+100 = 0.$$

Problem 4.* *Given that α, β, γ are the roots of the cubic equation $x^3-x^2+5x-3 = 0$, find the equation whose roots are $\beta+\gamma$, $\gamma+\alpha$, $\alpha+\beta$.*

Solution. By hypothesis,
$$\alpha+\beta+\gamma = 1, \qquad \beta\gamma+\gamma\alpha+\alpha\beta = 5, \qquad \alpha\beta\gamma = 3.$$
$$\therefore \beta+\gamma = 1-\alpha, \qquad \gamma+\alpha = 1-\beta, \qquad \alpha+\beta = 1-\gamma$$
and we require the equation whose roots† are $1-\alpha$, $1-\beta$, $1-\gamma$.

† We could work with $\beta+\gamma$ instead of $1-\alpha$, and so for the other roots; but the work would be more onerous.

Now $\qquad 1-\alpha+1-\beta+1-\gamma = 3-(\alpha+\beta+\gamma) = 2. \qquad (1)$

Again, $\qquad (1-\beta)(1-\gamma) = 1-(\beta+\gamma)+\beta\gamma,$

and so

$$(1-\beta)(1-\gamma)+(1-\gamma)(1-\alpha)+(1-\alpha)(1-\beta)$$
$$= 3-2(\alpha+\beta+\gamma)+\beta\gamma+\gamma\alpha+\alpha\beta$$
$$= 3-2+5 = 6. \qquad (2)$$

Finally,

$$(1-\alpha)(1-\beta)(1-\gamma) = 1-(\alpha+\beta+\gamma)+(\beta\gamma+\gamma\alpha+\alpha\beta)-\alpha\beta\gamma$$
$$= 1-1+5-3 = 2. \qquad (3)$$

Hence, from (1), (2), and (3), the required equation is, by the converse of Theorem 9,

$$x^3-2x^2+6x-2 = 0.$$

2.4. *Alternative method of solution.* The previous method is not always the easiest and the following method is well worth mastering. As a general rule, it is safe to say that the method now to be given should be used whenever it is applicable. On the other hand, the method of § 2.3 will suffice for all elementary examples.

Let α and β be the roots of the equation

$$ax^2+bx+c = 0. \qquad (1)$$

In (1) put $y = x+1$, i.e. $x = y-1$. The equation becomes

$$a(y-1)^2+b(y-1)+c = 0,$$

i.e. $\qquad ay^2+(b-2a)y+a-b+c = 0. \qquad (2)$

The roots of the equation (2) are $\alpha+1$ and $\beta+1$; for the values of y that satisfy (2) are connected with the values of x that satisfy (1) by means of the formula $y = x+1$, and the two values of x that satisfy (1) are α and β.

Hence, in order to find the equation whose roots are $\alpha+1$ and $\beta+1$ we put $y = x+1$ in (1) and write the result of so doing as a quadratic equation in y. The same method applies when y is any function of x; for example,

(i) the equation whose roots are $\frac{1}{2}\alpha$, $\frac{1}{2}\beta$ is obtained by putting $y = \frac{1}{2}x$, i.e. $x = 2y$: the result is $4ay^2+2by+c = 0$;

(ii) the equation whose roots are α^2, β^2 is obtained by putting $y = x^2$ in $ax^2+bx+c = 0$. This gives

$$ay+c = -bx,$$

and so
$$(ay+c)^2 = b^2x^2 = b^2y.$$

Thus the required equation is

$$a^2y^2+y(2ac-b^2)+c^2 = 0.$$

2.5. *Alternative solutions to Problems* 3 *and* 4.

PROBLEM 3. *Given that α and β are the roots of the equation $x^2+5x+10 = 0$, construct the equation whose roots are $2\alpha-3\beta$ and $2\beta-3\alpha$.*

SOLUTION. $\alpha+\beta = -5, \quad \alpha\beta = 10.$

Thus
$$2\alpha-3\beta = 5\alpha-3(\alpha+\beta) = 5\alpha+15,$$
and
$$2\beta-3\alpha = 5\beta-3(\alpha+\beta) = 5\beta+15.$$

In $x^2+5x+10 = 0$ put $y = 5x+15$. We get

$$\tfrac{1}{25}(y-15)^2+y-15+10 = 0,$$

i.e.
$$y^2-30y+225+25(y-5) = 0,$$
i.e.
$$y^2-5y+100 = 0.$$

The roots of this equation are $5\alpha+15$ and $5\beta+15$; that is, $2\alpha-3\beta$ and $2\beta-3\alpha$.

PROBLEM 4. *Given that α, β, γ are the roots of the cubic equation $x^3-x^2+5x-3 = 0$, find the equation whose roots are $\beta+\gamma$, $\gamma+\alpha$, $\alpha+\beta$.*

SOLUTION. $\alpha+\beta+\gamma = 1.$

We therefore require the equation whose roots are $1-\alpha$, $1-\beta$, $1-\gamma$. In $x^3-x^2+5x-3 = 0$ put $y = 1-x$, i.e. $x = 1-y$. We get

$$(1-y)^3-(1-y)^2+5(1-y)-3 = 0;$$

i.e.
$$2-6y+2y^2-y^3 = 0,$$

which is the equation required.

EXAMPLES III A

1. The roots of the equation $x^2-5x+10 = 0$ are α and β. Find the values of $\alpha^2+\beta^2$, $\alpha^3+\beta^3$, $\alpha^2\beta+\alpha\beta^2$.

2. The roots of the equation $x^2-7x+9 = 0$ are α and β. Find the values of $(\alpha-\beta)^2$ and $\alpha^4+\beta^4$.

HINT. $(\alpha-\beta)^2 = (\alpha+\beta)^2-4\alpha\beta$; $\alpha^4+\beta^4 = (\alpha^2+\beta^2)^2-2\alpha^2\beta^2$.

3. Given that α and β are the roots of $x^2-5x+3=0$, form the equation whose roots are

(i) α^3 and β^3; (ii) $3+\alpha$ and $3+\beta$; (iii) $1/\alpha$ and $1/\beta$;
(iv) $4\alpha+3\beta$ and $3\beta+4\alpha$.

HINT. (i) By § 2.3; (ii)–(iv) either by § 2.3 or § 2.4, preferably the latter.

4. Prove that, if α and β are the roots of the equation $ax^2+bx+c=0$, then

$$\alpha^2+\beta^2=\frac{b^2-2ac}{a^2}, \qquad \alpha^4+\beta^4=\frac{(b^2-2ac)^2-2a^2c^2}{a^4}.$$

5. Given that α and β are the roots of the equation $ax^2+bx+c=0$, construct the equation whose roots are

(i) $1/\alpha$ and $1/\beta$; (ii) $\alpha+1$ and $\beta+1$; (iii) 2α and 2β;
(iv) α^2 and β^2; (v) α^3 and β^3; (vi) $\alpha^2\beta$ and $\alpha\beta^2$.

6. One root of the equation $ax^2+bx+c=0$ is the square of the other root. Prove that $b^3=ca(3b-c-a)$.

7. The two roots of the equation $ax^2+bx+c=0$ differ by 5; prove that $b^2=4ac+25a^2$.

8. α and β are the roots of the equation $2x^2+4x+1=0$; prove that $\alpha^4+\beta^4=8\frac{1}{2}$ and that the roots of the equation $x^2-34x+1=0$ are α^2/β^2 and β^2/α^2.

9. Prove that, if α, β, γ are the roots of the equation

$$x^3+3x^2+2x+1=0,$$

(i) 2α, 2β, 2γ are the roots of $x^3+6x^2+8x+8=0$;
(ii) $1/\alpha$, $1/\beta$, $1/\gamma$ are the roots of $x^3+2x^2+3x+1=0$;
(iii) α^2, β^2, γ^2 are the roots of $y(y+2)^2=(3y+1)^2$.

10.* If α and β are the roots of the equation $2x^2-3x+4=0$, prove that the equation whose roots are $\alpha^2-\beta$ and $\beta^2-\alpha$ is $8x^2+26x+93=0$.
Prove that $\alpha^5+\beta^5=123/32$.

HINT. $\alpha^5+\beta^5=(\alpha^2+\beta^2)(\alpha^3+\beta^3)-\alpha^2\beta^2(\alpha+\beta)$.

11.* Given that α, β, γ are the roots of the cubic equation

$$x^3-x^2+5x-3=0,$$

find the values of

$$\alpha^2+\beta^2+\gamma^2, \quad \alpha^3+\beta^3+\gamma^3, \quad \alpha^{-1}+\beta^{-1}+\gamma^{-1}, \quad \beta^2\gamma^2+\gamma^2\alpha^2+\alpha^2\beta^2.$$

HINT. $\alpha^2+\beta^2+\gamma^2=(\alpha+\beta+\gamma)^2-2(\beta\gamma+\gamma\alpha+\alpha\beta)$;

$$\alpha^3+\beta^3+\gamma^3-3\alpha\beta\gamma=(\alpha+\beta+\gamma)(\alpha^2+\beta^2+\gamma^2-\beta\gamma-\gamma\alpha-\alpha\beta).$$

$$\beta^2\gamma^2+\gamma^2\alpha^2+\alpha^2\beta^2=(\beta\gamma+\gamma\alpha+\alpha\beta)^2-2\alpha\beta\gamma(\alpha+\beta+\gamma).$$

12. Given that α, β, γ are the roots of the cubic equation

$$x^3-2x^2+4x+7=0,$$

find the equation whose roots are

(i) $1/\alpha, 1/\beta, 1/\gamma$; (ii) $1+\alpha, 1+\beta, 1+\gamma$;

(iii) $\beta+\gamma, \gamma+\alpha, \alpha+\beta$; (iv) $\alpha^2, \beta^2, \gamma^2$.

13. The equation $x^3+3Hx+G=0$ has two equal roots. Prove that $G^2+4H^3=0$.

HINT. Let the roots be α, α, β, and use Theorem 9.

14. The roots of the equation $x^3+3Hx+G=0$ are in arithmetical progression. Prove that $G=0$.

HINT. Let the roots be $\alpha-\delta, \alpha, \alpha+\delta$; or use $\alpha+\gamma=2\beta$.

15.* Prove that, when $a_0 \neq 0$, the transformation $y=x+(a_1/a_0)$ changes the equation $a_0x^3+3a_1x^2+3a_2x+a_3=0$ into an equation of the form $a_0^3y^3+3a_0Hy+G=0$, where $G=2a_1^3-3a_0a_1a_2+a_0^2a_3$.

Deduce that, when the roots of the equation in x are in arithmetical progression, $2a_1^3-3a_0a_1a_2+a_0^2a_3=0$.

3. Formal solution of quadratic equations

3.1. Let a, b, c be given constants, $a \neq 0$. Let

$$ax^2+bx+c=0. \tag{1}$$

Then
$$x^2+\frac{b}{a}x=-\frac{c}{a},$$

and
$$x^2+\frac{b}{a}x+\left(\frac{b}{2a}\right)^2=\frac{b^2}{4a^2}-\frac{c}{a}.$$

That is,
$$\left(x+\frac{b}{2a}\right)^2=\frac{b^2-4ac}{4a^2}. \tag{2}$$

Hence
$$x+\frac{b}{2a}=\pm\frac{\sqrt{(b^2-4ac)}}{2a},$$

and so
$$x=\frac{-b\pm\sqrt{(b^2-4ac)}}{2a}. \tag{3}$$

When b^2-4ac is positive, we may write $b^2-4ac=N^2$, where N denotes the positive square root of b^2-4ac. The two roots of equation (1) are then

$$\frac{1}{2a}(-b+N) \quad \text{and} \quad \frac{1}{2a}(-b-N).$$

If $b^2-4ac = 0$, the equation (1) may be written, as we see from (2), in the form $\left(x+\dfrac{b}{2a}\right)^2 = 0$. We may, if we wish, say that the equation has only one root, namely, $-b/2a$; but it is more convenient, for the purposes of later applications, to say that the equation has two equal roots, or a double root (cf. § 1.1).

When b^2-4ac is negative, there is no ordinary elementary number whose square is b^2-4ac; for the square of such a number is always positive. We can either agree to say that no value of x can be found to satisfy (1), or invent a new type of number whose square shall be equal to the negative number b^2-4ac.

3.2. *Introduction of complex numbers.* For a long time everyone adopted the first alternative: they agreed to say that the equation had no root when b^2-4ac was negative. But this brought many difficulties in its train and, at length, mathematicians adopted the second alternative: they invented a type of number whose square would be a negative number.

The simplest procedure for inventing such a number is to say boldly let there be a number whose square is -1; let it be denoted† by the letter i; and let it be supposed that this symbol combines with other symbols such as x, a, b,... according to the usual rules of algebra; for example,

$$x+i = i+x,$$
$$x\times i = i\times x,$$
$$(a+b)i = ai+bi,$$

and so on.

On this procedure we make the HYPOTHESIS

'*There is a number, denoted by i, such that $i^2 = -1$, and if a is any other number, $(ai)^2 = -a^2$*';

we distinguish between the types of number thus:

(i) the numbers of ordinary elementary work we call REAL numbers; they may be positive, negative, or zero;

† In fact, it was the Greek letter ι, 'iota'; but the book-writers and printers afterwards preferred the English letter. In physics $\sqrt{(-1)}$ is often denoted by j; this because the letter i is then the accepted symbol for the electrical current.

(ii) if a is a real number, we call ai an IMAGINARY number;

(iii) if a and b are real numbers, we call $a+bi$ a COMPLEX number.

We make the further HYPOTHESIS

'*i can be combined with other symbols according to the ordinary laws of algebra.*'

3.3. *Nature of the roots of a quadratic equation.*

With the number i at our disposal the roots of the quadratic equation can be found even when b^2-4ac is negative. Suppose b^2-4ac is negative, and let M be the positive square root of $4ac-b^2$. Then, by (2),

$$\left(x+\frac{b}{2a}\right)^2 = -\frac{M^2}{4a^2} = \frac{M^2}{4a^2}\times(-1).$$

Hence
$$x+\frac{b}{2a} = \pm\frac{Mi}{2a}$$

and the two roots are

$$\frac{-b+Mi}{2a} \quad \text{and} \quad \frac{-b-Mi}{2a},$$

where $M^2 = 4ac-b^2$.

Thus, in all cases, the roots of the equation $ax^2+bx+c = 0$ are given by the formula

$$x = \frac{-b\pm\sqrt{(b^2-4ac)}}{2a}.$$

It is convenient at this point to state the results as a formal theorem.

THEOREM 10. *The equation $ax^2+bx+c = 0$, where a, b, c are real numbers, has two roots. Further,*

(i) *when b^2-4ac is positive, the two roots are real and different;*

(ii) *when b^2-4ac is zero, the two roots are equal, both being $-b/2a$, which is a real number;*

(iii) *when b^2-4ac is negative, the two roots are complex numbers.*

4. The definition of complex numbers

The logical foundations of the method adopted in § 3.2, where we simply made hypotheses that would enable us to write

$\sqrt{(-1)} = i$, $i^2 = -1$, and to combine i with other symbols, are open to some criticism. It is best to make a fresh start and define complex numbers more carefully than we did in § 3.2. This we shall do in Volume II.

Meanwhile, we note the following facts:

(1) the algebra of complex numbers is precisely the same as that of ordinary (real) numbers, save that i^2 is replaced by -1 whenever it occurs;

(2) two complex numbers $a+bi$ and $c+di$, where a, b, c, d are real numbers, are equal if and only if $a = c$ and $b = d$;

(3) the number $a-bi$ is called the CONJUGATE COMPLEX of the number $a+bi$;

(4) the number $+\sqrt{(a^2+b^2)}$ is called the MODULUS of the complex number $a+bi$; it is denoted by $|a+bi|$. Also $(a-bi)(a+bi) = a^2+b^2$, which is the square of the modulus.

(5) a complex number may, if convenient, be denoted by a single symbol, such as z, Z, a, p, or any other letter. If z denotes the complex number $x+yi$ (where x and y are real numbers) we refer to x as the REAL PART of z, and to y as the IMAGINARY PART of z. The notation $|z|$ then denotes the modulus, $+\sqrt{(x^2+y^2)}$, of the complex number.

(6) When, by using (2) above, we deduce the two facts $a = c$ and $b = d$ from the equation $a+bi = c+di$, the procedure is commonly referred to as 'equating real and imaginary parts'.

5. Calculations involving complex numbers

We work out the details of a few simple problems that involve the use of complex numbers.

PROBLEM 1. *Prove that* (i) *the sum,* (ii) *the difference,* (iii) *the product, and* (iv) *the quotient, of the two complex numbers* $3+i$ *and* $2-3i$ *is itself a complex number.*

SOLUTION. (i) *Sum* $(3+i)+(2-3i)$

$$= 3+i+2-3i = 5-2i.$$

(ii) *Difference.*

$$(3+i)-(2-3i) = 3+i-2+3i = 1+4i.$$

(iii) *Product.*

$$(3+i)(2-3i) = 6-9i+2i-3i^2$$
$$= 6-7i+3 \quad (i^2 = -1)$$
$$= 9-7i.$$

(iv) *Quotient.*†

$$\frac{3+i}{2-3i} = \frac{(3+i)(2+3i)}{(2-3i)(2+3i)}$$
$$= \frac{6+11i-3}{4+9}$$
$$= \frac{3}{13}+\frac{11}{13}i.$$

NOTE. The same results hold for the sum, difference, product, and quotient of any two complex numbers. In particular

$$(a+bi)(c+di) = (ac-bd)+i(bc+ad),$$

a result that follows at once on writing $bdi^2 = -bd$, and

$$\frac{a+bi}{c+di} = \frac{(a+bi)(c-di)}{(c+di)(c-di)}$$
$$= \frac{(ac+bd)+i(bc-ad)}{c^2+d^2}$$
$$= \frac{ac+bd}{c^2+d^2}+i\frac{bc-ad}{c^2+d^2}.$$

PROBLEM 2. *Find the quadratic equation whose roots are $2+3i$ and $2-3i$.*

SOLUTION. The equation is $x^2+px+q = 0$, where

$$p = -(2+3i+2-3i) = -4,$$

and
$$q = (2+3i)(2-3i) = 4+9 = 13.$$

The equation is $\quad x^2-4x+13 = 0.$

PROBLEM 3. *Prove that* $(1+i)^2 = 2i$, $(1+i)^3 = -2(1-i)$, $(1+i)^4 = -4$.

† Notice that $(2-3i)(2+3i) = 4-9i^2 = 4+9$.

SOLUTION.
$$(1+i)^2 = 1+2i+i^2 = 1+2i-1 = 2i;$$
$$(1+i)^3 = 1+3i+3i^2+i^3 = 1+3i-3-i = -2+2i;$$
$$(1+i)^4 = 1+4i+6i^2+4i^3+i^4 = 1+4i-6-4i+1 = -4.$$

NOTE. The values of i^2, i^3, i^4, i^5,... are given by $i^2 = -1$; $i^3 = i^2.i = -i$; $i^4 = -i.i = +1$; $i^5 = i$;.... That is, $i = i$, $i^2 = -1$, $i^3 = -i$, $i^4 = +1$, $i^5 = +i$, $i^6 = -1$, $i^7 = -i$, and so on, the sequence i, -1, $-i$, $+1$ repeating itself.

PROBLEM 4.* *Prove that the three cube roots of unity are* 1, ω, *and* ω^2, *where*
$$\omega = \cos\frac{2\pi}{3}+i\sin\frac{2\pi}{3} = \frac{-1+i\sqrt{3}}{2},$$
and that
$$1+\omega+\omega^2 = 0.$$

SOLUTION. Let z be a number whose cube is equal to unity. That is,
$$z^3 = 1.$$
Then
$$z^3-1 = 0, \tag{1}$$
i.e.
$$(z-1)(z^2+z+1) = 0. \tag{2}$$
Thus,† either $z = 1$, or $z^2+z+1 = 0$.

The two roots of the equation $z^2+z+1 = 0$ are given by
$$z = \frac{-1\pm\sqrt{(-3)}}{2}.$$
On using the facts
$$\cos\frac{2\pi}{3} = -\frac{1}{2},\quad \sin\frac{2\pi}{3} = \frac{\sqrt{3}}{2},\quad \cos\frac{4\pi}{3} = -\frac{1}{2},\quad \sin\frac{4\pi}{3} = -\frac{\sqrt{3}}{2},$$
these two roots may be written as
$$\cos\frac{2\pi}{3}+i\sin\frac{2\pi}{3} \quad\text{and}\quad \cos\frac{4\pi}{3}+i\sin\frac{4\pi}{3}.$$
Moreover, denoting the first of these by ω, we have
$$\omega^2 = \left(\cos\frac{2\pi}{3}+i\sin\frac{2\pi}{3}\right)^2$$
$$= \left(\cos^2\frac{2\pi}{3}-\sin^2\frac{2\pi}{3}\right)+2i\sin\frac{2\pi}{3}\cos\frac{2\pi}{3}$$
$$= \cos\frac{4\pi}{3}+i\sin\frac{4\pi}{3},$$

† When $a\times b = 0$, either (i) $a = 0$ or (ii) $b = 0$ or (iii) both a and b are zero.

so that the two roots are ω and ω^2, and the three roots of $z^3 = 1$ are 1, ω, and ω^2.

Finally, since ω is a value of z that satisfies the equation $1+z+z^2 = 0$, $1+\omega+\omega^2 = 0$.

NOTE. The reader who intends to read Volume II should lay no stress on the present solution of Problem 4. The problem is at once simpler and more illuminating when the later theory has been mastered.

PROBLEM 5.* *Find real numbers a and b such that*

$$(a+bi)^2 = 3+4i.$$

SOLUTION. If $(a+bi)^2 = 3+4i$, then

$$a^2+2abi-b^2 = 3+4i.$$

On equating real and imaginary parts (cf. (2) of § 4),

$$a^2-b^2 = 3, \qquad 2ab = 4. \tag{1}$$

The simplest method of solving the equations (1) is to note that $a = 2$, $b = 1$ is an obvious solution, whence $a = 2$, $b = 1$ is one solution and $a = -2$, $b = -1$ is another.

Assuming we have not been quick enough to see the obvious solution at a glance, we can proceed thus:

From the second equation of (1), $b = 2/a$, whence, on substitution in the first equation,

$$a^4-3a^2-4 = 0,$$

i.e. $\qquad (a^2-4)(a^2+1) = 0, \qquad a^2 = 4 \text{ or } -1.$

We want the *real* number a. We therefore discard the solution $a^2 = -1$ and obtain $a = \pm 2$.

When $a = 2$, the equation $b = 2/a$ gives $b = 1$;
When $a = -2$, the equation $b = 2/a$ gives $b = -1$.
The solution is $a = 2$, $b = 1$ or $a = -2$, $b = -1$.

EXAMPLES III B

1. Prove that

(i) $(3+i)+(7-6i) = 10-5i = 5(2-i)$,

(ii) $2i-3+7-5i = 4-3i$, (iii) $(3+i)(7-6i) = 27-11i$,

(iv) $(2i-5)(4i-3) = 7-26i$, (v) $(2-i)(3-i) = 5(1-i)$.

2. Prove that $i^3 = -i$, and that

$$(3i^3+5)(2i^3+3) = 9-19i,$$
$$(5-3i^3)(3-2i^3) = 9+19i.$$

3. Prove, by the method of § 5, Problem 1, *Quotient*, that

(i) $\dfrac{3+i}{2+i} = \dfrac{7-i}{5}$, $\dfrac{3-i}{2-i} = \dfrac{7+i}{5}$;

(ii) $\dfrac{6-5i}{3-2i} = \dfrac{28-3i}{13}$, $\dfrac{6+5i}{3+2i} = \dfrac{28+3i}{13}$;

(iii) $\dfrac{6-5i}{3+2i} = \dfrac{8-27i}{13}$, $\dfrac{6+5i}{3-2i} = \dfrac{8+27i}{13}$.

4. Prove that

(i) $|2+3i| = \sqrt{13}$, $|2-3i| = \sqrt{13}$, $(2+3i)(2-3i) = 13$;

(ii) $|3+4i| = 5$, $|3-4i| = 5$, $(3+4i)(3-4i) = 25$;

(iii) $|5+i| = \sqrt{26}$, $|5-i| = \sqrt{26}$, $(5+i)(5-i) = 26$.

5. By means of the formula (3) of § 3.1, find the roots of the quadratic equations [do not work out $\sqrt{37}$, etc.]

(i) x^2-7x+3, (ii) $x^2-7x+13$,

(iii) x^2-5x+2, (iv) x^2-5x+7.

6. Find the quadratic equations whose roots are

(i) $2+\sqrt{3}$ and $2-\sqrt{3}$, (ii) $2+3i$ and $2-3i$,

(iii) $3+\sqrt{5}$ and $3-\sqrt{5}$, (iv) $3+i\sqrt{5}$ and $3-i\sqrt{5}$,

(v) $\frac{1}{2}(1+\sqrt{2})$ and $\frac{1}{2}(1-\sqrt{2})$, (vi) $\frac{1}{2}(1+i\sqrt{3})$ and $\frac{1}{2}(1-i\sqrt{3})$.

7. Prove that

$$(1+i)(2+i)(3+i) = 10i,$$
$$(1+i)(1+2i)(1+3i) = -10.$$

8. Prove that

$$\frac{1+i}{1-i} = i, \quad \frac{2+i}{3-i} = \frac{1+i}{2}, \quad \frac{2+i}{2-i} = \frac{3+4i}{5}.$$

9. Prove that $(1+i)^6 = -8i$.

HINT. Find $(1+i)^2$ and cube the result.

10.* The complex numbers Z and z, where $Z = X+Yi$, $z = x+yi$, and X, Y, x, y are real numbers, are connected by the formula

$$z = \frac{Z+1}{Z-1}.$$

Prove that

$$x+yi = \frac{X^2+Y^2-1-2Yi}{(X-1)^2+Y^2},$$

and deduce that

$$x = \frac{X^2+Y^2-1}{(X-1)^2+Y^2}, \qquad y = \frac{-2Y}{(X-1)^2+Y^2}.$$

11.* Prove that, when ω is a complex cube root of unity,

$$(a+b\omega+c\omega^2)(a+b\omega^2+c\omega) = a^2+b^2+c^2-bc-ca-ab,$$
$$(a+b+c)(a+b\omega+c\omega^2)(a+b\omega^2+c\omega) = a^3+b^3+c^3-3abc.$$

12.** Prove that, when ω is a complex cube root of unity and

$$x = a+b,\ y = a+b\omega,\ z = a+b\omega^2,$$
$$|y-z| = |z-x| = |x-y|.$$

6. The coefficients of polynomials restricted to be real

Although we have introduced complex numbers in order to find the roots of such equations as $x^2+1 = 0$ or $x^2+2x+3 = 0$, we shall not in this book consider such equations as

$$(2+i)x^2+(3+4i)x-(2-i) = 0,$$

or such polynomials as

$$(3+i)x^3-(2+5i)x^2-3x+4i.$$

It is only on rare occasions that polynomials having complex numbers as *coefficients* come under consideration; and in this book we shall exclude them altogether.

The coefficients of all polynomials that occur in this book are to be taken as real numbers. THE FACT THAT a, b, c,... APPEAR IN THE TEXT AS COEFFICIENTS OF A POLYNOMIAL WILL IMPLY THAT a, b, c,... ARE REAL NUMBERS; no explicit statement that a, b, c,... are real will be made unless the fact requires particular stress. The alternative would be a constant and tiresome reiteration of such phrases as 'where a and b are real numbers'.

Thus, although the *roots* of an equation may sometimes be complex numbers, the *coefficients* (as far as we shall be concerned) will not be complex numbers.

CHAPTER IV

QUADRATIC FORMS

1. The sign of a product of factors

In this section a, b, c are given numbers, a is less than b, and b is less than c; in symbols $a < b < c$.

1.1. *The sign of $x-a$.* When x is greater than a, $x-a$ is positive; and when x is less than a, $x-a$ is negative.

1.2. *The sign of $(x-a)(x-b)$.*

(i) When x is less than a, both $x-a$ and $x-b$ are negative; their product is positive.

(ii) When x lies between a and b, x is greater than a but less than b; the factor $x-a$ is positive, while the factor $x-b$ is negative; their product is negative.

(iii) When x is greater than b, both factors are positive; their product is positive.

As x increases from values less than a to values greater than b, there are two points at which the product changes sign, namely, $x = a$ and $x = b$; at these points the product is equal to zero.

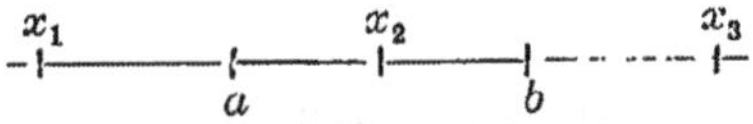

x_1 — a — x_2 — b — x_3

NOTE. The diagram on the right, where x_1 represents (i) above, x_2 and x_3 represent (ii) and (iii), is useful; it brings out the signs of the factors. For example, x_2 lies to the right of a and to the left of b; so that x_2-a is positive and x_2-b is negative.

1.3. *The sign of $(x-a)(x-b)(x-c)$; when $a < b < c$.* If x is less than a, all three factors are negative and the product is negative.

If x is greater than a but less than b and c, one factor is positive and two factors are negative; the product is positive.

The reader will see for himself that when x lies between b and c, the product is negative; and that when x is greater than c, the product is positive.

THE SIGN OF A PRODUCT OF TWO, THREE, OR MORE FACTORS IS OFTEN AN IMPORTANT DETAIL.

1.4. *The sign of* $(x-a)^2$ *and of* $(x-a)^3$. The square $(x-a)^2$ is zero when $x = a$ and is positive for all other real values of x; for whether $x-a$ is positive or negative, its square is positive.

The cube $(x-a)^3$ is negative when $x < a$, is zero when $x = a$, and is positive when $x > a$.

2.* The positive-definite quadratic form

2.1. The expression

$$ax^2+2bx+c \quad (a \neq 0),$$

where a, b, c are real numbers, is commonly called a QUADRATIC FORM. The particular quadratic forms whose values are always positive when x is real have a special importance in many branches of mathematics.

We begin by noting that a quadratic form which has real factors does NOT possess this property. For, if

$$ax^2+2bx+c \equiv a(x-\alpha)(x-\beta),$$

where α and β are real, $(x-\alpha)(x-\beta)$ takes negative values when x lies between α and β and takes the value zero when $x = \alpha$ or β; and if

$$ax^2+2bx+c \equiv a(x-\alpha)^2,$$

where α is real, its value is zero (which is not positive) when $x = \alpha$.

With these preliminary remarks made, we proceed to the formal proofs of our theorems.

2.2. THEOREM 11 A. *If the quadratic form*†

$$ax^2+2bx+c \quad (a \neq 0)$$

is positive for all real values of x, *then*

$$a > 0, \quad c > 0, \quad and \quad ac > b^2.$$

PROOF. Let $ax^2+2bx+c > 0$ for all real values of x. Then there is no real value of x for which $ax^2+2bx+c = 0$. Hence $b^2-ac < 0$ (Theorem 10; with $2b$ instead of b), i.e. $ac > b^2$.

When $x = 0$ the value of the form is c, which is therefore positive.

† We exclude the case $a = 0$ because we are discussing a *quadratic* form; and the form would not be quadratic if a were zero.

Finally, since $ac > b^2$ and c is positive, a must also be positive.

2.3. THEOREM 11 B. *If the coefficients a, b, c satisfy the conditions*
$$a > 0, \quad c > 0, \quad \textit{and} \quad ac > b^2,$$
then the quadratic form $\quad ax^2+2bx+c$
takes positive values for all real values of x.

PROOF. Let $a > 0$, $c > 0$, and $ac > b^2$. Then†

$$ax^2+2bx+c \equiv a\left\{\left(x+\frac{b}{a}\right)^2+\frac{c}{a}-\frac{b^2}{a^2}\right\}$$
$$\equiv a\left\{\left(x+\frac{b}{a}\right)^2+\frac{ac-b^2}{a^2}\right\}. \tag{1}$$

When x is real, the term $(x+b/a)^2$ is either positive ($x \neq -b/a$) or zero ($x = -b/a$), while, by hypothesis, the term $(ac-b^2)/a^2$ is positive. Hence the sum of the terms within the 'curly brackets' is positive whenever x is real. Moreover, $a > 0$, by hypothesis, so that the product (1) is positive.

It follows that $ax^2+2bx+c$ is positive whenever x is real.‡

Before proceeding we sum up the results of Theorems 11 A and 11 B in one theorem.

THEOREM 11. *The necessary and sufficient conditions that the quadratic form $ax^2+2bx+c$ should be positive for all real values of x are*
$$a > 0, \quad c > 0, \quad ac > b^2.$$

Such a form is called a POSITIVE-DEFINITE FORM.

2.4. *'Necessary and sufficient conditions.'* We have just used, for the first time in this book, a phrase that often occurs in algebra. Consider the two statements,

(A) the quadratic form $ax^2+2bx+c$ is positive for all real values of x;

(B) $a > 0$, $c > 0$, and $ac > b^2$.

† The steps are merely those of 'completing the square'.

‡ It may be noted that we have used only *two* of the conditions, $a > 0$ and $ac > b^2$. But when these two conditions hold, then also $c > 0$; for if c were less than or equal to zero while a was positive, the product ac would be negative or zero, and the condition $ac > b^2$ could not be satisfied.

Equally, if $c > 0$ and $ac > b^2$, then also $a > 0$. It is thus simpler to state the theorem with all three conditions as hypotheses.

If we take at random any set of values of the constants a, b, c, we cannot expect to find these statements true—they may or may not be true. What we proved in § 2.2 was that whenever (A) is true, (B) is true also; in other words, (B) is a NECESSARY consequence of (A). What we proved in § 2.3 was that the truth of (B) is SUFFICIENT to ensure the truth of (A). Thus, the truth of (B) is both a necessary consequence of the truth of (A) and a sufficient hypothesis for the truth of (A) to be ensured. In such a case we say that (B) is a set of necessary and sufficient conditions for (A).

The analogous detail in geometry is 'the theorem and its converse'. For example, the theorem 'the angles at the base of an isosceles triangle are equal' has as its converse 'if two angles of a triangle are equal, the triangle is isosceles'. If this were a theorem in algebra, it would probably be enunciated in the form 'the necessary and sufficient condition for two angles of a triangle to be equal is that two sides should be equal'. The equality of two sides is necessary—for whenever two angles are equal the equality of the sides necessarily follows; moreover, the equality of the sides is a sufficient hypothesis for a proof that two angles are equal.

The reader will have learnt long ago that a theorem is not the same as its converse and that a proof of the converse is not a proof of the theorem itself. He should learn to exercise in algebra the same care that he has learnt to exercise in geometry, and, when proving that (A) is a necessary and sufficient condition for (B), he should be quite sure that he proves the two propositions

(i) if (B) is true, then (A) is true;

(ii) if (A) is true, then (B) is true.

3.* Other quadratic forms

3.1. When $ac > b^2$, but both a and c are negative, the quadratic form $ax^2+2bx+c$ is NEGATIVE for all real values of x.

PROOF.

$$ax^2+2bx+c \equiv a\left\{\left(x+\frac{b}{a}\right)^2+\frac{ac-b^2}{a^2}\right\}, \qquad (1)$$

and, as in § 2.3, the expression within the curly brackets is positive for *all* real values of x. But a is now negative, and so (1) is negative. Such a form is called a NEGATIVE-DEFINITE form.

3.2. When $ac < b^2$, the roots of the equation

$$ax^2+2bx+c = 0$$

are two distinct real numbers α and β, say. By what we have seen in § 1.2, $ax^2+2bx+c \equiv a(x-\alpha)(x-\beta)$ is positive for some values of x, negative for others, and zero when $x = \alpha, \beta$.

When $ac = b^2$, the quadratic $ax^2+2bx+c$ may be written as $(x-\alpha)^2$, where α is real. This is positive for all real values of x save $x = \alpha$, when it is zero.

3.3. Notice that

(i) when the roots of $ax^2+2bx+c = 0$ are real and different, the value of $ax^2+2bx+c$ may be positive, negative, or zero, according to the particular value x may have;

(ii) when the roots of $ax^2+2bx+c = 0$ are real and coincident ($b^2 = ac$), and $a > 0$, the value of $ax^2+2bx+c$ may be positive ($x \neq -b/a$) or zero ($x = -b/a$);

(iii) when the roots of $ax^2+2bx+c = 0$ are not real, but are complex, the value of $ax^2+2bx+c$ is always positive if a is positive, and is always negative if a is negative.

EXAMPLES IV

1. Prove that x^2-3x+2 is positive when $x < 1$ and when $x > 2$, but is negative when $1 < x < 2$.

2. Prove that $x^3-6x^2+11x-6$ is negative when $x < 1$, and is positive when $x > 3$. For what other values of x is it positive?

[ANS. $1 < x < 2$.]

3. Prove that $4x^2+5x-6$ is positive when $x < -2$, negative when $-2 < x < \frac{3}{4}$, positive when $x > \frac{3}{4}$.

4. Prove that

(i) $3x^2+8x+5$ is negative when $-\frac{5}{3} < x < -1$;

(ii) $3x^2+8x-3$ is positive when $x < -3$, and when $x > \frac{1}{3}$;

(iii) $4x^2+15x+14$ is positive when $x < -2$.

5. Find the factors of $4x^3+19x^2+29x+14$ and show that the expression is negative when $x < -2$, positive when $-2 < x < -\frac{7}{4}$, negative when $-\frac{7}{4} < x < -1$, and positive when $x > -1$.

6.* Prove that $x^2+6x+17$, $x^2-10x+42$, x^2-4x+5 are positive for all real values of x.

7.* Prove that $4x^2-4x+3$, $9x^2+12x+5$, $2x^2-6x+5$ are positive for all real values of x.

8.* Prove that $3x-2x^2-2$ is negative for all real values of x.

9. Prove that $(x^2+2)(x^2-1)$ is positive when $x > 1$ and when $x < -1$, but is negative when $-1 < x < 1$.

10. The equation $x^2+bx+c = 0$ has two real roots α and β. Prove that $x^2+bx+c < 0$ when x lies between α and β.

CHAPTER V

POLYNOMIALS: GRAPHS

1. Use of the differential calculus

It is assumed† that the reader is already familiar with the procedure of plotting the numerical values of one variable, y, against the numerical values of another variable, x, and so obtaining 'the graph of $y = f(x)$'.

It is further assumed† that the reader can differentiate simple functions of x and is familiar with the following facts:

(i) when (dy/dx) is positive, y increases as x increases and the graph of $y = f(x)$ slopes from 'bottom left' to 'top right', as in Fig. 1;

FIG. 1 FIG. 2

(ii) when (dy/dx) is negative, y decreases as x increases and the graph of $y = f(x)$ slopes from 'top left' to 'bottom right', as in Fig. 2;

(iii) if $(dy/dx) = 0$ when $x = a$, and the sign of dy/dx changes from positive to negative as x increases from values just

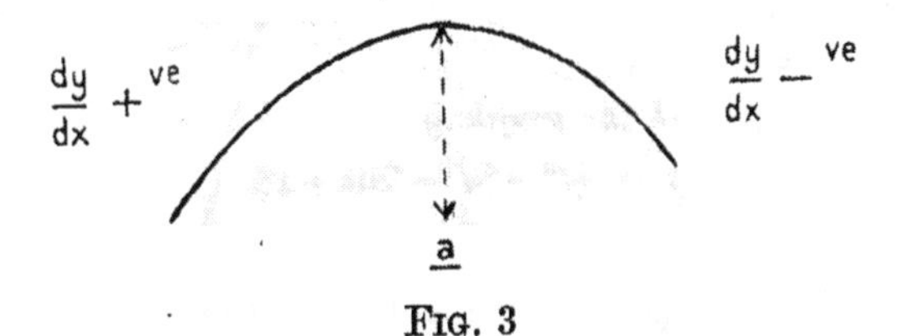

FIG. 3

below a to values just above a, there is a maximum at $x = a$, as in Fig. 3;

† If these assumptions are not correct, the reader should defer the reading of Chapter V until such time as they are correct.

(iv) if $(dy/dx) = 0$ when $x = a$, and the sign of (dy/dx) changes from negative to positive as x increases from values just below a to values just above a, there is a minimum at $x = a$, as in Fig. 4;

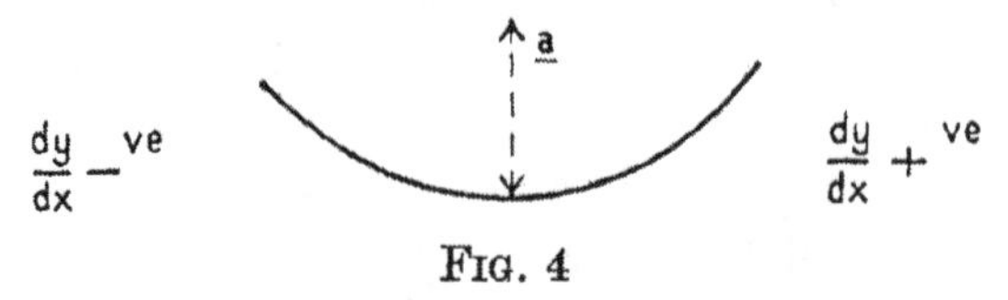

FIG. 4

(v) a point that is either a maximum or a minimum on the curve is commonly called a 'turning-point'.

2. The graphs of particular polynomials

In studying polynomials, and other functions, it is often useful to be able to form an idea of the general shape of the graph, without recourse to the plotting of a very large number of points. To form this idea it is sufficient to determine the shape of the graph near its 'turning-points' and 'points of inflexion' (if any), and to fix one or two 'control points' which, when plotted, show the run of the graph (a) between the turning-points and (b) beyond the turning-points.

It is not a question of plotting carefully a bit of the graph between $x = -4$ and $x = +4$, say, but of finding the general shape of the graph as x varies from large negative values, through zero, to large positive values. The process is called 'sketching the graph' or 'drawing a rough graph'. We illustrate it by examples.

EXAMPLE 1. *Sketch the graph of*

$$y = 3x^5+5x^3-30x+10.$$

SOLUTION.

$$\frac{dy}{dx} = 15(x^4+x^2-2) = 15(x^2+2)(x^2-1)$$
$$= 15(x^2+2)(x+1)(x-1). \tag{1}$$

This is zero when $x = -1$, $x = +1$, or $x^2+2 = 0$, the last being impossible for real values of x. When $x = -1$, $y = 32$ and when $x = 1$, $y = -12$, and so the possible turning-points are $(-1, 32)$ and $(1, -12)$.

To find whether these points are maxima or minima we consider the change in sign of (1) as x passes through the values -1 and $+1$. Now

x^2+2 is positive for all real values of x;

$x+1$ is negative when $x < -1$, positive when $x > -1$;

$x-1$ is negative when $x < 1$, positive when $x > 1$.

Thus $\qquad -1 \qquad +1$

(a) dy/dx is positive to the left of -1; it is $(+)(-)(-)$;†

(b) dy/dx is negative between -1 and $+1$; it is $(+)(+)(-)$;

(c) dy/dx is positive to the right of $+1$; it is $(+)(+)(+)$.

Accordingly, as x increases through the value -1, (dy/dx) changes from positive to negative and there is a maximum at $x = -1$. Also, as x increases through the value $+1$, (dy/dx) changes from negative to positive and there is a minimum at $x = +1$.

To sketch the graph we first mark in the parts of the graph

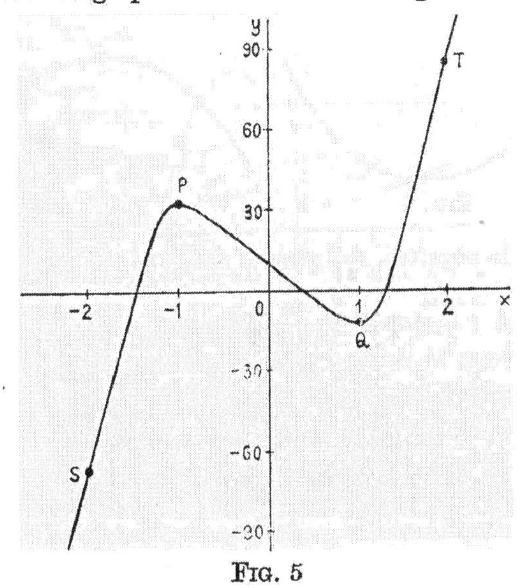

FIG. 5

† This notation means that when $x < -1$, (dy/dx) is the product of factors whose signs are $+$, $-$, $-$; and so for (b) and (c).

near the turning-points (P and Q of the figure) making sufficient curve to show the maximum at P and the minimum at Q.

Next we plot three 'control-points', taking one value of x less than -1, one between -1 and $+1$, and one greater than $+1$.

$$x = -2, \quad 0, \quad 2,$$
$$y = -66, \quad 10, \quad 86.$$

When these points are marked in we can complete the part PQ of the graph and see that, beyond the turning-points, the curve goes off in the directions PS and QT of the figure.

3. The graph of a quadratic form

3.1. When $y = ax^2+2bx+c \quad (a \neq 0)$,

$$\frac{dy}{dx} = 2(ax+b)$$

and the graph has one turning-point, $x = -b/a$.

When a is positive, the general shape of the graph is shown by Fig. 6.

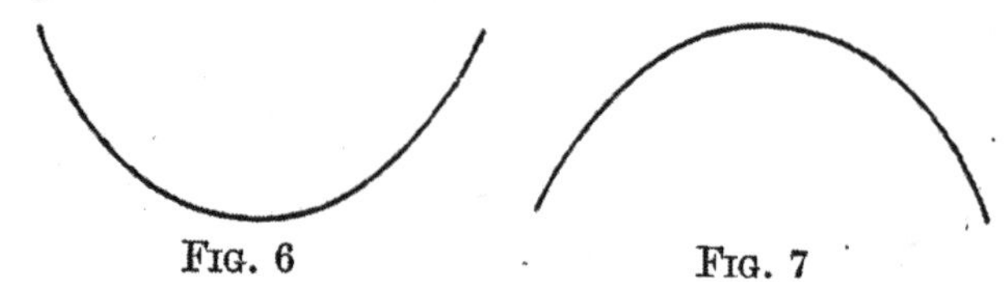

FIG. 6 FIG. 7

When a is negative, the general shape of the graph is shown by Fig. 7.

Examples 1 and 2 at the end of the chapter will give the reader exercise in dealing with particular cases. In working these examples it is sufficient (i) to find the turning-point, (ii) to plot a control-point on each side of the turning-point.

4. The graph of a cubic; typical examples

EXAMPLE 2. *Sketch the graph of* $y = x^3-3x+4$.

SOLUTION.

$$\frac{dy}{dx} = 3x^2-3 = 3(x+1)(x-1). \tag{1}$$

The possible turning-values are -1 and $+1$.

To find whether these points are maxima or minima we consider the sign of (1). Now

$x+1$ is negative to the left of -1, positive to the right of -1;

$x-1$ is negative to the left of $+1$, positive to the right of $+1$.

Hence† $\quad -1 \quad\quad +1$

(a) (dy/dx) is positive when $x < -1$; it is $(-)(-)$;

(b) (dy/dx) is negative when $-1 < x < 1$; it is $(+)(-)$;

(c) (dy/dx) is positive when $x > 1$; it is $(+)(+)$.

There is a maximum at $x = -1$ and a minimum at $x = +1$.

As control-points we plot

$$x = -2, \quad 0, \quad 2,$$
$$y = \quad 2, \quad 4, \quad 6,$$

and obtain the curve as shown.

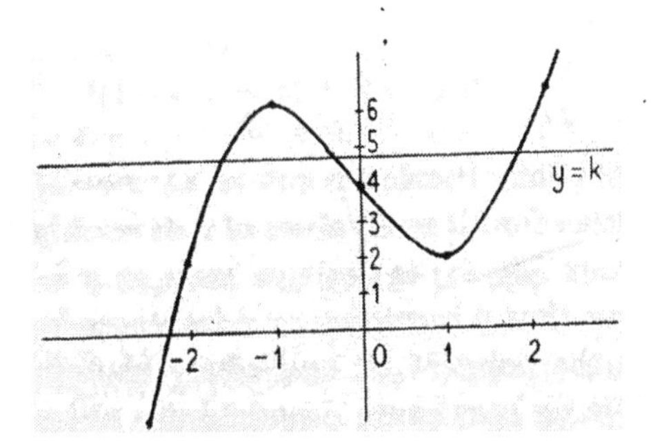

Fig. 8

Deductions from the graph. The roots of the equation

$$x^3-3x+4 = 0 \tag{1}$$

are, by definition, the values of x for which x^3-3x+4 is zero, that is, the values of x for which y is zero. By inspection of the graph, there is only one real value of x (roughly, $-2\frac{1}{4}$) for which y is zero.

† See footnote on Example 1, p. 51.

In Chapter III, § 1, we saw that a cubic equation has three roots: it follows that the remaining two roots of the equation (1) are not real numbers, but complex numbers.

Hence the equation (1) has one real and two complex roots.

Again, if k is a number *between* 2 and 6, the roots of the equation $x^3-3x+4-k = 0$ are given by those points on the graph at which $y = k$. By inspection of the graph, the equation $x^3-3x+4-k = 0$ has three real roots.

The line $y = 2$ touches the graph; the equation

$$x^3-3x+4-2 = 0$$

may be considered to have two equal roots at $x = 1$ and another at $x = -2$. In fact,

$$\begin{aligned} x^3-3x+4-2 &\equiv x^3-3x+2 \\ &\equiv (x-1)^2(x+2), \end{aligned}$$

so that the *two equal roots correspond to the squared factor.*

EXAMPLE 3. *Sketch the graph of* $y = x^3-3x^2+3x+2$.

SOLUTION. $\dfrac{dy}{dx} = 3(x^2-2x+1) = 3(x-1)^2.$

The only possible turning-point is at $x = 1$, $y = 3$. But $(x-1)^2$ is positive for all real values of x save only $x = 1$, when it is zero. Hence (dy/dx) is positive save at $x = 1$, when it is zero. It follows that y increases as x increases both before the curve reaches the point (1, 3) and after the curve leaves that point. There is no maximum or minimum at $x = 1$; there is a POINT OF INFLEXION, the run of the curve being as shown.

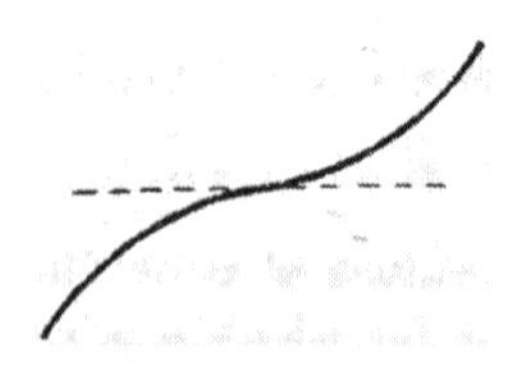

As control-points we plot

$$x = 0,\ y = 2;\ x = 2,\ y = 4;\ x = -1,\ y = -5.$$

The curve is then as shown.

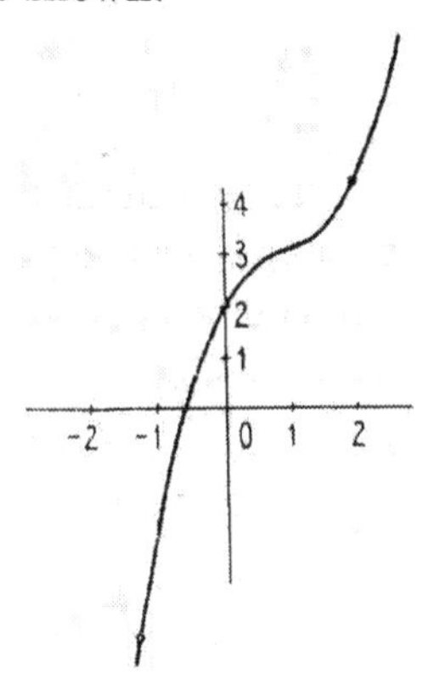

FIG. 9

DEDUCTIONS FROM THE GRAPH:

(1) The equation $x^3-3x^2+3x+2 = 0$ has only one real root, given by the point on the graph at which $y = 0$, a value lying between 0 and -1.

(2) The real roots of the equation $x^3-3x^2+3x+2-k = 0$ are given by the points where the graph meets the line $y = k$. For all values of k other than 3 there is only one real root.

The line $y = 3$ touches the graph: it may be considered to meet it in three coincident points. In fact, when $k = 3$, $x^3-3x^2+3x+2-k$ reduces to x^3-3x^2+3x-1, which is $(x-1)^3$; and we regard the equation $(x-1)^3 = 0$ as having three equal roots 1, 1, 1.

EXAMPLE 4. *Sketch the graph of* $y = x^3+3x-2$.

SOLUTION. $$\frac{dy}{dx} = 3x^2+3 = 3(x^2+1).$$

Now x^2+1 is positive for *all* real values of x: it is never zero. Hence y increases with x at all points of the graph.

In order to sketch such a curve with as little actual plotting as possible, we must find the point at which d^2y/dx^2 is zero.

This gives the point of inflexion† on the curve. In the present example

$$\frac{d^2y}{dx^2} = 6x,$$

which is zero when $x = 0$. The value of dy/dx when $x = 0$ is 3; this gives the slope of the curve at the point of inflexion.

To check the run of the curve on either side of the point of inflexion we plot the control-points

$$x = -1,\ y = -6, \qquad x = 1,\ y = 2.$$

The curve is as shown.

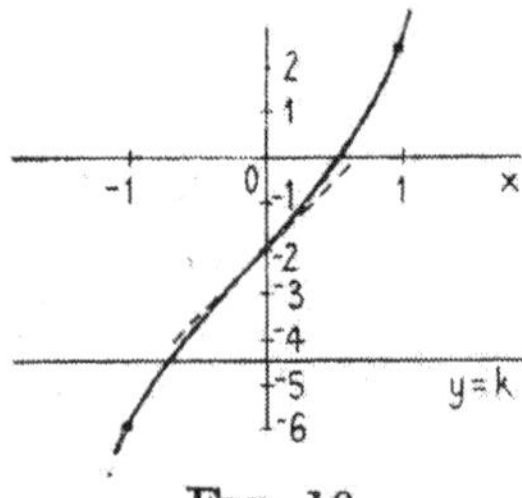

Fig. 10

In drawing the curve, first plot the point of inflexion $(0, -2)$, then draw sufficient of the curve through this point to mark the fact that the slope of the curve at the point is 3, and, finally, mark in the control-points and fill in the graph. The dotted line is the tangent at $(0, -2)$; at a point of inflexion the curve crosses the tangent.

Deduction from the graph. The equation

$$x^3+3x-2-k = 0$$

has only one real root for all values of k. It is given by the point where the graph meets the line $y = k$. The figure shows a negative value of k.

5.* A general property of cubic equations

Theorem 12. *Every cubic equation with real coefficients p, q, r, s either has one real root and two complex roots or has three real roots.*

† This is a result proved by differential calculus.

Let the cubic equation be

$$px^3+qx^2+rx+s = 0 \quad (p \neq 0).$$

Then, on dividing by p, and making an obvious change of notation for the coefficients, the equation may be written as

$$x^3+3ax^2+3bx+c = 0. \tag{1}$$

[The factors 3 are brought in because they simplify the resulting arithmetic.]

The graph $y = x^3+3ax^2+3bx+c$ cuts Ox $[y = 0]$ at the points given by the real roots of (1).

Now $$\frac{dy}{dx} = 3(x^2+2ax+b),$$

and $$\frac{d^2y}{dx^2} = 6(x+a).$$

Also $$x^2+2ax+b = 0 \tag{2}$$

when $$(x+a)^2 = a^2-b,$$

i.e. $$x = -a\pm\sqrt{(a^2-b)}.$$

(i) If $a^2 > b$, the roots of (2) are real and distinct. Let them be α and β, where $\alpha < \beta$. Then

$$\frac{dy}{dx} = 3(x-\alpha)(x-\beta). \tag{3}$$

When $x < \alpha$, (3) is positive; when $\alpha < x < \beta$, (3) is negative; when $x > \beta$, (3) is positive. There is a maximum at $x = \alpha$ and a minimum at $x = \beta$. The general shape of the curve is

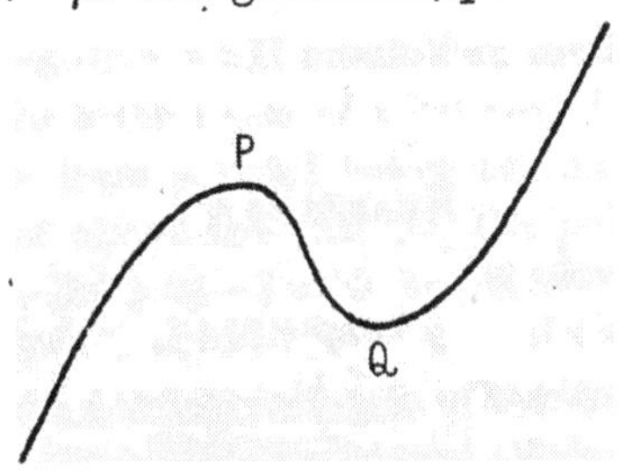

FIG. 11

If the axis Ox lies above P or below Q, it cuts the graph at one point only; if the axis Ox is at the level of P (or of Q) it touches the graph at P (or Q) and cuts it at one other point; if the axis Ox lies below P and above Q it cuts the graph at three distinct points.

Hence the equation (1) has only one real root when Ox lies above P or below Q; in all other cases it has three real roots, two equal and one other when Ox is at the level of P or Q, and three distinct real roots when Ox falls between the levels of P and Q.

(ii) If $a^2 = b$,

$$\frac{dy}{dx} = 3(x+a)^2,$$

$$\frac{d^2y}{dx^2} = 6(x+a),$$

and the shape of the graph is that of Fig. 9, § 4.

If the axis Ox falls either above or below the point of inflexion, it cuts the graph in one point only. If Ox is the tangent to the graph at the point of inflexion, it may be considered to meet the curve in three coincident points there.†

(iii) If $a^2 < b$, the graph has no real turning-point. Moreover, by Theorem 11, dy/dx is positive for all values of x; and the general shape of the graph is that of Fig. 10, § 4. The axis Ox cuts such a graph in one point only.

NOTE ON THEOREM 12. The result given is a particular case of the theorem that every equation of degree n either has n real roots or has an even number of complex roots. We shall prove this general theorem in Volume II.

EXAMPLES V A

1. Sketch the graphs of

$$y = 2x^2+3x+1, \qquad y = 3x^2+2x+1, \qquad y = x^2-2x-4.$$

2. Sketch the graphs of

$$y = 3-2x-3x^2, \qquad y = 3-4x-2x^2, \qquad y = 6+x-x^2.$$

† The curve is $y = x^3+3ax^2+3a^2x+c$; the point of inflexion is $(-a, c-a^3)$, and the tangent at this point is $y = c-a^3$. This tangent is the axis Ox when $c = a^3$ and then the curve is $y = (x+a)^3$ so that the values of x for which y is zero are given by $(x+a)^3 = 0$.

3. Sketch the graphs of

(i) $y = 2x^3-3x^2-12x+7$, (ii) $y = -2x^3+3x^2+12x-7$,

(iii) $y = 7+12x-x^3$, (iv) $y = x^3-12x-7$.

4. Sketch the graph of $y = 2x^3-9x^2+12x-5$, and find, from your graph, the values of k for which the equation $2x^3-9x^2+12x-5 = k$ has two equal roots.

5. Sketch the graph of $y = 4x^3-6x^2+3x-1$, and show that the equation $4x^3-6x^2+3x-1 = c$ has only one real root unless $c = -\frac{1}{2}$.

6. Sketch the graph of $y = 7-12x+6x^2-x^3$.

7. Sketch the graph of $y = 2x^3+12x+9$.

8. Sketch the graph of each of the following:

(i) $y = x^4-8x^3+22x^2-24x+9$,

(ii) $y = x^5-5x+7$,

(iii) $y = 3x^5-15x^4+40x^3-60x^2+45x+7$.

9. (*An example in which the numbers do not 'work out' to whole numbers or easy fractions.*) Sketch the graph of

$$y = 3x^3+7x^2-6x+5.$$

6. Numerical approximations to the roots of an equation

6.1. If a polynomial changes sign between the two values $x = a$ and $x = b$, it must be equal to zero for some value of x between a and b. For example, when

$$y = x^3-3x^2+4x-3,$$

the table of values

$x =$	0,	1,	2,	3,
$y =$	-3,	-1,	1,	9,

shows that y is negative when $x = 1$ and positive when $x = 2$. It must be zero for some value of x between 1 and 2, since the graph of y passes from a point below the axis Ox, the point $(1, -1)$, to a point above the axis Ox, the point $(2, 1)$. Hence the equation $x^3-3x^2+4x-3 = 0$ has a root lying between $x = 1$ and $x = 2$.

In order to make a closer estimate of the numerical value of this root we may proceed in any one of three ways. We may

(i) draw the graph of y on a fairly large scale for values of x between 1 and 2, and read off the value of x at which the graph cuts the x-axis;

(ii) test a few values of x between 1 and 2, e.g.

$$x = \quad 1{\cdot}5, \qquad 1{\cdot}6, \qquad 1{\cdot}7,$$
$$y = -0{\cdot}375, \quad -0{\cdot}184, \quad 0{\cdot}043,$$

and deduce that the root lies between 1·6 and 1·7, probably very near the latter since 0·043 is quite small;

(iii) use the method of successive approximations given in the next section.

6.2. *Newton's method of approximation.* Let $f(x)$ denote any polynomial in x, and let $f'(x)$ denote its differential coefficient. Then

$$\frac{f(a+h)-f(a)}{h} \to f'(a) \qquad \text{as } h \to 0.$$

It follows that, when h is small,

$$f(a+h)-f(a) \simeq hf'(a),$$

or

$$f(a+h) \simeq f(a)+hf'(a), \tag{1}$$

where $\simeq$ means 'is approximately equal to'.

This approximation for $f(a+h)$ is fundamental in all branches of mathematics. We shall apply it here to the problem of finding the numerical value of a root of an equation.

Let $$y = f(x) \equiv x^3-3x^2+4x-3,$$
so that $$f'(x) \equiv 3x^2-6x+4.$$

Our previous table of values (§ 6.1) gives

$$x = \quad 0, \quad 1, \quad 2, \quad 3,$$
$$y = -3, \quad -1, \quad 1, \quad 9,$$

and this shows that $y = 0$ for a value of x somewhere near $x = \frac{3}{2}$ (y is as much above Ox at $x = 2$ as it is below Ox at $x = 1$, and so crosses Ox somewhere near the point half-way between 1 and 2). Suppose the actual value is $\frac{3}{2}+h$, so that

$$f(\tfrac{3}{2}+h) = 0.$$

By (1), $$f(\tfrac{3}{2})+hf'(\tfrac{3}{2}) \simeq 0,$$
i.e. $$-\tfrac{3}{8}+\tfrac{7}{4}h \simeq 0,$$
or $$h \simeq \tfrac{12}{56} \simeq 0{\cdot}2.$$

Thus a first approximation to the root is $1{\cdot}5+0{\cdot}2 = 1{\cdot}7$.

6.3. *Second approximation.* In order to obtain a better approximation, we merely repeat the process using 1·7 instead of 1·5. We know there is a root near 1·7; suppose it is $1{\cdot}7+k$, so that

$$f(1{\cdot}7+k) = 0.$$

Then, by (1), $$f(1{\cdot}7)+kf'(1{\cdot}7) \simeq 0,$$

and so, on working out the values of $f(1{\cdot}7)$ and $f'(1{\cdot}7)$,

$$0{\cdot}043+k(2{\cdot}47) \simeq 0.$$

This gives $$k \simeq -\frac{0{\cdot}043}{2{\cdot}47} \simeq -0{\cdot}02,$$

so that the root is approximately equal to 1·68.

The process can be repeated to give closer and closer approximations, but the work soon becomes burdensome.

NOTES. (i) The first approximation is sometimes very rough; it is not necessarily accurate to the first place of decimals and if we are to be at all certain that our approximation has that degree of accuracy, we must check it by obtaining at least a rough estimate of the next approximation.

(ii) If the values of y at $x = 1$ and at $x = 2$, say, have opposite signs but the former is small in comparison with the latter, the root lies near $x = 1$ and we put $x = 1+h$ to find the first approximation.

(iii) When using this method avoid using

$$f(a+h) \simeq f(a)+hf'(a)$$

for a value of a which makes $f'(a)$ small: to find h we have to *divide* by $f'(a)$ and so it is better to have $f'(a)$ large rather than small.

EXAMPLES V B

1. Show that each of the equations

$$\text{(i)}\quad x^7-5x+3 = 0,$$
$$\text{(ii)}\quad x^5+3x^2-8 = 0,$$
$$\text{(iii)}\quad x^7-60x-1 = 0,$$

has a root between $x = 1$ and $x = 2$, and obtain an approximation to this root correct to one place of decimals.

2. Show that each of the equations

$$\text{(i)}\quad x^3+7x-3 = 0,$$
$$\text{(ii)}\quad x^3-7x+3 = 0,$$
$$\text{(iii)}\quad x^5+4x^2-1 = 0,$$

has a root between $x = 0$ and $x = 1$, and obtain an approximation to this root correct to one place of decimals.

3. Find correct to one place of decimals, the roots of the equation $70x^3-81x^2-100x+96 = 0$.

HINT. A table of values will show that the roots lie, one between -2 and -1, one between 0 and 1, one between 1 and 2.

4. Prove that the equation

$$14x^4+23x^3-16x^2+23x-30 = 0$$

has one root between 0 and 1 and one between -3 and -2. Find these roots, correct to one place of decimals.

5. Show that the only real root of the equation

$$x^5-10x+15 = 0$$

is approximately equal to $-2{\cdot}04$.

HINT. The only change of sign occurs between -2 and -3.

CHAPTER VI

PRODUCT OF FACTORS; THE BINOMIAL THEOREM

1. The product $(x+a)(x+b)...(x+k)$

1.1. We know, from elementary work on 'factors', that

$$(x+a)(x+b) \equiv x^2+(a+b)x+ab. \qquad (1)$$

This elementary result is easily proved by actually multiplying out the two factors on the left; some readers will do it 'mentally', others will set out the multiplication sum in full.

When we multiply by a third factor $x+c$, we get

$$(x+a)(x+b)(x+c) \equiv x^3+(a+b+c)x^2+(bc+ca+ab)x+abc. \quad (2)$$

This could be proved by setting out in full the multiplication of (1) by $x+c$. But this is tedious, and it is not necessary; and we get the result another way.

We think of x as a variable, and we think of a, b, c as constants. To find the product of the three factors on the left of (2) we want the sum of all the products that arise by multiplying together one term from each bracket. [Compare (1), where R.H.S. is $x^2+ax+bx+ab$.]

(i) We may take x from each of the brackets, when the product of the three x's is x^3.

(ii) We may take the x from two of the brackets and the constant from the remaining bracket; we can do this in three different ways, for we may take the constant either from the first, second, or third bracket. The three terms we get in this way are ax^2, bx^2, cx^2.

(iii) We may take the x from one bracket and the constant from the other two brackets; we can do this in three different ways. The three terms we get in this way are bcx, cax, abx.

(iv) We may take the constant from each bracket, when the product of the three constants is abc.

Thus the sum total of terms obtained on multiplying the three factors $(x+a)(x+b)(x+c)$ is

$$x^3+(a+b+c)x^2+(bc+ca+ab)x+abc,$$

which proves (2).

NOTE. The reader should teach himself, by a thorough understanding of (i)–(iv) above, to write down (2) without any explanation.

1.2. A particular case of (2) is given by taking $b = c = a$. The identity (2) then becomes

$$(x+a)^3 \equiv x^3+3ax^2+3a^2x+a^3, \tag{3}$$

a result which the reader probably knows already.

1.3. We can repeat the procedure of §§ 1.1 and 1.2 with four factors $x+a$, $x+b$, $x+c$, $x+d$, and, on putting all constants equal to a, obtain

$$(x+a)^4 \equiv x^4+4ax^3+6a^2x^2+4a^3x+a^4. \tag{4}$$

But already the details are a little troublesome to write down, and when we try it for five or six factors they become extremely tiresome. The problem we want to solve is 'What corresponds to (3) and (4) when we have $(x+a)^n$, where n is any positive integer?'

In § 1.1 it was easy enough to see that in writing down the product in (2) there were three terms containing x^2, namely, x^2a, x^2b, and x^2c; and three containing x, namely, xbc, xca, and xab. Before we can discuss $(x+a)^n$ we must consider the question 'in how many ways can we choose x from one, two, three,... of the n brackets

$$(x+a)(x+b)(x+c)...(x+k),$$

and take the constants from the remaining brackets?' This question we broach in §§ 2 and 3.

2. Permutations

2.1. Suppose we are given 3 letters a, b, c. We can write them down in 6 different orders, namely,

$$abc,\ acb;\ bac,\ bca;\ cab,\ cba. \tag{5}$$

We examine these orders carefully as a guide to the general problem when there are r letters to be arranged in order. We can choose any one of the 3 letters a, b, c to fill the first place: there are thus 3 ways of filling the first place.

When we have filled the first place, we are left with 2 letters, of which we may choose either for the second place; thus 2 ways of filling the second place are associated with each way of filling the first place, and so there are 3×2 ways of filling the first two places.

When we have filled the first two places, we are left with only 1 letter and it must go in the last place: thus there are $3\times 2\times 1$ ways of filling the three places.

In (5) we have written down to begin with the two orders in which a comes first, then the two orders in which b comes first, and then the two orders in which c comes first.

2.2. *The factorial notation.* The number $3\times 2\times 1$ is denoted by $3!$; the number $4\times 3\times 2\times 1$ is denoted by $4!$; and so on. When n is any positive integer,

$$n! \quad \text{denotes} \quad n(n-1)(n-2)\ldots 1. \qquad (6)$$

We refer to the symbol $n!$ as 'factorial n', though some people refer to it (with solemn faces) as 'n shriek'.

2.3. THEOREM 13. *The r different letters $a, b,\ldots, k$ (say) can be arranged in $r!$ different orders.*

PROOF. We may choose any one of the r letters for the first place in the order. Having filled the first place in any particular way, we may choose any one of the remaining $r-1$ letters for the second place; thus $r-1$ different ways of filling the second place are associated with each of the r ways of filling the first place. Hence there are $r(r-1)$ different ways of filling the first two places in the order.

Having filled the first two places in any one particular way, we can fill the third in $r-2$ different ways; hence there are $r(r-1)(r-2)$ different ways of filling the first three places.

Continuing thus, we see that there are

$$r(r-1)(r-2)\ldots 2.1$$

different ways of filling the r places, and so we can arrange the letters in $r!$ different orders.

PERMUTATIONS. *We refer to the different orders as the permutations of the r letters.*

2.4. THEOREM 14. *Given r places, numbered $1, 2, \ldots, r$, and n letters ($n \geqslant r$), there are*

$$n(n-1)\ldots(n-r+1)$$

different ways of assigning r of the letters to the r numbered places.

PROOF. We can fill the first place in n different ways. Having filled the first place in any one particular way, we are left with $n-1$ letters and so can fill the second place in $n-1$ different ways. Hence there are $n(n-1)$ different ways of filling the first two places.

Having filled the first two places in any particular way, we can fill the third in $n-2$ different ways; so that we can fill the first three places in $n(n-1)(n-2)$ different ways. Continuing thus, we see that the number of ways of filling the r places is

$$n(n-1)(n-2)\ldots \text{ to } r \text{ factors.}$$

Now the *second* factor is $n-1$, the *third* is $n-2$, and the rth factor is $n-(r-1)$, i.e. $n-r+1$. Hence the number in question is

$$n(n-1)(n-2)\ldots(n-r+1).$$

2.5. We sometimes refer to the $n(n-1)\ldots(n-r+1)$ ways of filling the r numbered places as the 'number of permutations of the n letters taken r at a time'. This number is also denoted by ${}_nP_r$. Thus, when $n > r$,

$${}_nP_r \equiv n(n-1)(n-2)\ldots(n-r+1) = \frac{n!}{(n-r)!}; \tag{7}$$

also (Theorem 13), $\quad {}_nP_n = n!$

3. Combinations

3.1. In Theorem 14 we considered the problem 'Given n letters, in how many different ways can we choose r of them to fill r numbered places?' In this problem the arrangement $abcd\ldots$, wherein a occupies first place and b occupies second place, is counted as distinct from the arrangement of the same letters in the order $bacd\ldots$, wherein b comes first and a comes second.

We now consider the problem 'Given n letters, in how many

different ways can we choose r of them, no regard being paid to the order in which we choose them'. A simple example will serve to illustrate the difference between the two problems.

Suppose we are given 3 letters a, b, c.

FIRST PROBLEM. *To fill two places numbered* 1 *and* 2. We can do this in 3×2 ways; for we can fill place 'No. 1' in 3 ways, using either a, b, or c to do so, and having filled place 'No. 1' in any one of these 3 ways, we can then fill place 'No. 2' in 2 ways. The 6 ways of filling the 2 places are

$$a, b;\ a, c;\ b, a;\ b, c;\ c, a;\ c, b. \tag{8}$$

SECOND PROBLEM. *To choose* 2 *out of the* 3 *letters, no regard being paid to the order of choice.* We can do this in 3 ways; for we can choose (i) b and c, or (ii) c and a, or (iii) a and b; and there is no other way of choosing 2 letters.

The relation between the two problems is also easy to see in this simple example.

In (8) we reckon as distinct the two arrangements (i) first a and then b, (ii) first b and then a. That is, corresponding to the one choice 'a and b' in the second problem, there are 2 different arrangements in the first problem.

Accordingly, the answer '6' to the first problem is twice the answer '3' to the second problem.

3.2. THEOREM 15. *Given n letters, the number of different ways in which we can choose r of them ($n > r$), no regard being paid to the order of choice, is*

$$\frac{n(n-1)...(n-r+1)}{r!}. \tag{9}$$

PROOF. Suppose that the number of ways in which we can choose r letters out of the n, no regard being paid to order of choice, is X_r. Consider any one selection; it consists of r letters. These r letters can (Theorem 13) be arranged in $r!$ different orders. Hence, given n letters, there are $X_r \times (r!)$ ways in which we can fill r places, numbered 1, 2,..., r. But (Theorem 14) the number of ways in which we can do this is

$$n(n-1)...(n-r+1).$$

Hence $$r! \times X_r = n(n-1)...(n-r+1),$$
and the number X_r is given by the formula (9) above.

3.3. COMBINATIONS. *The number of ways in which r letters can be selected from n letters, no regard being paid to the order of choice, is called the number of combinations of n letters taken r at a time.*

This number is commonly denoted by $_nC_r$; an alternative notation is ^nC_r. By Theorem 15,

$$_nC_r \equiv \frac{n(n-1)...(n-r+1)}{r!} \equiv \frac{n!}{r!\,(n-r)!}. \tag{10}$$

EXAMPLES VI A

In 1–4 the multiplications should be done mentally, the coefficients of the various powers of x being obtained by the method explained in § 1.1.

1. Prove that

$$(x-a)(x-b) = x^2-(a+b)x+ab,$$
$$(x-a)(x-b)(x-c) = x^3-(a+b+c)x^2+(bc+ca+ab)x-abc.$$

2. Prove that

$$(x+1)(x+2)(x+3) = x^3+6x^2+11x+6,$$
$$(x-1)(x-2)(x-3) = x^3-6x^2+11x-6.$$

3. Prove that

$$(x+1)(x-2)(x+3) = x^3+2x^2-5x-6,$$
$$(x^2-1)(x+5) = x^3+5x^2-x-5,$$
$$(x-1)(x+2)(x+3) = x^3+4x^2+x-6.$$

4. Prove that

$$(x^2-1)(x^2-2)(x^2+3) = x^6-7x^2+6,$$
$$(ax+b)(ax+2b)(ax+3b) = a^3x^3+6a^2bx^2+11ab^2x+6b^3.$$

---

5. In how many ways can one

(i) pick 3 teams from a list of 10 teams?

(ii) pick 1 team for first place, 1 for second place, and 1 for third place, from a list of 10 teams?

6. A group of 22 people is to be divided into two groups of 11 each. In how many different ways can it be done?

HINT. The problem is that of choosing one group of 11; for when the first group is chosen, the other group must consist of those left out of the first group.

7. In how many different ways can

(i) 1 boy and 1 girl be chosen from a group of 6 boys and 9 girls?

(ii) 2 boys and 3 girls be chosen from a group of 6 boys and 9 girls?

HINT. With each way of choosing the boy(s) we can associate each way of choosing the girl(s): the answer to (i) is, accordingly, 6×9.

8. In how many ways can we choose

(i) 7 letters out of 10 letters, (ii) 4 letters out of 6 letters,

(iii) 5 letters out of 8 letters, (iv) 3 letters out of 8 letters,

no regard being paid to the order of choice?

9. In how many ways can we arrange in order 7 football teams; 4 letters; 5 examination candidates?

10. We are given 10 letters and 7 numbered places; in how many different ways can we fill the numbered places?

11. There are 7 vacancies, one in each of 7 different counties; there are 10 candidates in all. In how many ways can we fill the vacancies?

HINT. Number the counties and use Example 10.

12. A group of 10 people is to be divided into one group of 7 and one group of 3; in how many ways can it be done?

HINT. In Examples 10, 11 the *order* of choice is relevant; in Example 12 we have to select 7 people, no regard being paid to the order in which they are selected.

---

13. Work out the numerical values of

$$_5P_3,\quad {}_7P_5,\quad {}_5C_3,\quad {}_7C_5.$$

14. Prove that $_nC_4 > {}_nC_3$ if $n > 7$.

SOLUTION.

$$_nC_4 = \frac{n(n-1)(n-2)(n-3)}{4!} = \frac{n-3}{4}\times {}_nC_3.$$

Let $n > 7$; then $n-3 > 4$, and $_nC_4$ is equal to $_nC_3$ multiplied by a factor that is greater than unity. Hence $_nC_4 > {}_nC_3$.

15. Prove that

(i) $_nC_6 > {}_nC_5$ if $n > 11$,

(ii) $_nC_6 < {}_nC_5$ if $n < 11$,

(iii) $_nC_r > {}_nC_{r-1}$ if $n+1 > 2r$.

16. Verify that

(i) $_7C_4+{}_7C_3 = {}_8C_4$,

(ii) $_6C_3+{}_6C_2 = {}_7C_3$,

(iii) $_8C_6+2\,{}_8C_5+{}_8C_4 = {}_{10}C_6$.

4. The binomial theorem

4.1. THEOREM 16. *When n is a positive integer, $(x+a)^n$ is identically equal to*

$$x^n + {}_nC_1\, x^{n-1}a + \ldots + {}_nC_r\, x^{n-r}a^r + \ldots + a^n.$$

PROOF. Consider the product

$$(x+a)(x+a)\ldots(x+a) \quad [n \text{ brackets}].$$

It is (compare § 1) the sum of all the products we can obtain by multiplying together one term from each bracket. Further,

(i) we may take x from each of the brackets, when the product of the n x's is x^n;

(ii) we may take a from one bracket and an x from each of the remaining brackets, and we can do this in n ways; there are n products $x^{n-1}a$;

(iii) when r is an integer less than n, we can select r out of the n brackets in ${}_nC_r$ ways, and if, having chosen our r brackets, we take a out of each of them and take x out of the $n-r$ brackets we have not chosen, the product is $x^{n-r}a^r$, so that there are ${}_nC_r$ products $x^{n-r}a^r$;

(iv) we may take a from each of the n brackets, when the product is a^n.

Thus the sum of all the terms we get on multiplying out the n brackets is

$$x^n + {}_nC_1\, x^{n-1}a + \ldots + {}_nC_r\, x^{n-r}a^r + \ldots + a^n, \tag{11}$$

and this proves the theorem.

4.2. *Notes on Theorem* 16.

NOTE 1. If we think of $(a+x)^n$, instead of $(x+a)^n$, the result we have proved in (11) above shows that

$$(a+x)^n \equiv a^n + {}_nC_1\, a^{n-1}x + \ldots + {}_nC_r\, a^{n-r}x^r + \ldots + x^n. \tag{12}$$

But $(a+x)^n \equiv (x+a)^n$, and so (11) and (12) are identically equal: in (11) the coefficient of $x^r a^{n-r}$ is† ${}_nC_{n-r}$ and in (12) it is ${}_nC_r$, so that

$${}_nC_{n-r} = {}_nC_r. \tag{13}$$

This means that, in (11), coefficient of $x^{n-1}a$ = coefficient of xa^{n-1}, coefficient of $x^{n-2}a^2$ = coefficient of x^2a^{n-2}, and so on.

† Notice that, in (11), the suffix after C and the power of x add up to n.

The important result (13) is also obvious from the expressions for the two symbols given in (10). By (10),

$$ {}_nC_{n-r} \equiv \frac{n(n-1)...(r+1)}{(n-r)!} \equiv \frac{n!}{(n-r)!\,r!}, $$

and

$$ {}_nC_r \equiv \frac{n(n-1)...(n-r+1)}{r!} \equiv \frac{n!}{r!\,(n-r)!}. $$

NOTE 2. The coefficients

$$ 1,\quad {}_nC_1 \equiv n,\quad {}_nC_2 \equiv \frac{n(n-1)}{2!},\ldots,\quad {}_nC_r \equiv \frac{n(n-1)...(n-r+1)}{r!},\ldots $$

are often referred to as THE BINOMIAL COEFFICIENTS.

NOTE 3. The name 'binomial' is a survival of the old technical term; $x+y$ is a 'binomial', $x+y+z$ is a 'trinomial', while $x+y+...+\lambda$, where there are more than three letters, is a 'multinomial'.

NOTE 4. Sometimes it is more convenient to use the notation of Theorem 16 and to write

$$ (x+a)^n \equiv x^n + {}_nC_1 x^{n-1}a + ... + {}_nC_r x^{n-r}a^r + ... + a^n, $$

and sometimes it is more convenient to put in the values of the binomial coefficients and to write

$$ (x+a)^n \equiv x^n + nx^{n-1}a + \frac{n(n-1)}{2!}x^{n-2}a^2 + ... + $$
$$ + \frac{n(n-1)...(n-r+1)}{r!}x^{n-r}a^r + ... + a^n. \quad (14) $$

NOTE 5. *The general term.* The expression (11), or its equivalent, (14), is called the expansion of $(x+a)^n$ in descending powers of x; the expression (12) is called the expansion of $(x+a)^n$ in ascending powers of x.

The expression

$$ {}_nC_r x^{n-r}a^r \quad\text{or}\quad \frac{n(n-1)...(n-r+1)}{r!}x^{n-r}a^r $$

is called the general term of the expansion, to distinguish it from the particular terms, x^n, $nx^{n-1}a$, and so on.

4.3. *Numerical examples of Theorem* 16. In using the binomial theorem to expand expressions like $(1+x)^7$ or $(2+x)^6$, we avoid needless labour if we apply the result of (13). For example,

$$ {}_7C_1 = 7, \quad {}_7C_2 = \frac{7.6}{2} = 21, \quad {}_7C_3 = \frac{7.6.5}{3.2} = 35, $$

and so the expansion of $(1+x)^7$ begins

$$ 1+7x+21x^2+35x^3+\ldots. $$

But, by (13), the coefficient of x^4 = the coefficient of x^3,

" " " x^5 = " " " x^2,

" " " x^6 = " " " x,

and the coefficient of x^7 is 1. Hence

$$ (1+x)^7 = 1+7x+21x^2+35x^3+35x^4+21x^5+7x^6+x^7. $$

RULE. *When the index n is odd, calculate ${}_nC_r$ as far as $r = \frac{1}{2}(n-1)$, and then use the result of* (13).

Again, to expand $(2+x)^6$, we calculate

$$ {}_6C_1 = 6, \quad {}_6C_2 = \frac{6.5}{2} = 15, \quad {}_6C_3 = \frac{6.5.4}{3.2} = 20. $$

The expansion of $(2+x)^6$ begins

$$ 2^6+6.2^5.x+15.2^4.x^2+20.2^3.x^3+\ldots. $$

From (13) the coefficient of 2^2x^4 = the coefficient of 2^4x^2,

" " " $2x^5$ = " " " 2^5x,

and the coefficient of x^6 is 1. Hence

$$ \begin{aligned} (2+x)^6 &= 2^6+6.2^5x+15.2^4x^2+20.2^3x^3+15.2^2x^4+6.2x^5+x^6 \\ &= 64+192x+240x^2+160x^3+60x^4+12x^5+x^6. \end{aligned} $$

RULE. *When the index n is even, calculate ${}_nC_r$ as far as $r = \frac{1}{2}n$, and then use the result of* (13).

5. Particular examples of the binomial theorem

5.1. $(x+a)^3 = x^3+3x^2a+3xa^2+a^3,$ (15)

$(x+a)^4 = x^4+4x^3a+6x^2a^2+4xa^3+a^4,$ (16)

$(x+a)^5 = x^5+5x^4a+10x^3a^2+10x^2a^3+5xa^4+a^5.$ (17)

It will be noticed that in (15) and (17), which correspond to odd values of n, the numerical coefficients increase and then decrease, there being two equal coefficients in the middle; in (16), which corresponds to an even value of n, the numerical coefficients increase to a largest term in the middle and then decrease.

We shall return later to the general theorem on this point.

5.2. *Numerical approximations.* The binomial expansion can be used to find the value of expressions such as $(1\cdot002)^{13}$, $(1\cdot01)^7$ to any desired number of significant figures.

EXAMPLE. *Find the value of* $(1\cdot01)^5$ *to 4 significant figures.*

SOLUTION.

$$\begin{aligned}(1\cdot01)^5 &= (1+10^{-2})^5 \\ &= 1+5.10^{-2}+10.10^{-4}+10.10^{-6}+5.10^{-8}+10^{-10}.\end{aligned}$$

Write the values down column-wise, stopping as soon as it becomes clear that further entries in the column would not affect the first four figures of the answer.

$$\begin{aligned}1 &= 1 \\ 5.10^{-2} &= 0\cdot05 \\ 10.10^{-4} &= 0\cdot001 \\ 10.10^{-6} &= 0\cdot00001 \\ \text{Add} \quad & \quad 1\cdot051 \quad \text{(to 4 significant figs.)}\end{aligned}$$

The two terms we have left out, 5.10^{-8} and 10^{-10}, have no effect unless we want the answer to 7 places of decimals.

5.3. *The expansion of* $(x-a)^n$.

$$\begin{aligned}(x-a)^n &\equiv x^n+nx^{n-1}(-a)+\frac{n(n-1)}{2!}x^{n-2}(-a)^2+\ldots+ \\ &\quad +\frac{n(n-1)\ldots(n-r+1)}{r!}x^{n-r}(-a)^r+\ldots+(-a)^n \\ &\equiv x^n-nx^{n-1}a+\frac{n(n-1)}{2!}x^{n-2}a^2-\ldots+ \\ &\quad +(-1)^r\frac{n(n-1)\ldots(n-r+1)}{r!}x^{n-r}a^r+\ldots+(-1)^n a^n,\end{aligned}$$

and it is in the latter form that one is accustomed to use the expansion. For example,

$$(a-b)^5 = a^5-5a^4b+10a^3b^2-10a^2b^3+5ab^4-b^5.$$

5.4. In this section we work out one or two problems that are typical of the more elementary applications of the binomial theorem.

PROBLEM 1. *Expand* $\left(x+\frac{1}{x}\right)^6$ *in descending powers of* x.

SOLUTION.

$$\left(x+\frac{1}{x}\right)^6 = x^6+6x^5.\frac{1}{x}+15x^4.\frac{1}{x^2}+20x^3.\frac{1}{x^3}+15x^2.\frac{1}{x^4}+6x.\frac{1}{x^5}+\frac{1}{x^6}$$

$$= x^6+6x^4+15x^2+20+15x^{-2}+6x^{-4}+x^{-6}.$$

The same work gives, on a rearrangement of the terms,

$$\left(x+\frac{1}{x}\right)^6 = \left(x^6+\frac{1}{x^6}\right)+6\left(x^4+\frac{1}{x^4}\right)+15\left(x^2+\frac{1}{x^2}\right)+20,$$

which is a type of result that is often useful in the trigonometry associated with de Moivre's theorem.

PROBLEM 2. *Find the expansion of*

$$(2+x-3x^2)^7$$

in ascending powers of x, *calculating the coefficients as far as that of* x^4.

SOLUTION.† $(2+x-3x^2)^7 = \{2+x(1-3x)\}^7.$ (1)

The coefficients in the expansion of $(a+b)^7$ are

$$1, \quad 7, \quad \frac{7.6}{2} = 21, \quad \frac{7.6.5}{3.2} = 35, \quad \frac{7.6.5.4}{4.3.2} = 35, \ldots$$

and so (1) is equal to

$$2^7+7.2^6x(1-3x)+21.2^5x^2(1-3x)^2+35.2^4x^3(1-3x)^3+ \\ +35.2^3x^4(1-3x)^4+\ldots. \quad (2)$$

We now expand $(1-3x)^2$, $(1-3x)^3$, $(1-3x)^4$ but omit all terms that lead to powers of x above the fourth. We get

$$2^7+7.2^6(x-3x^2)+21.2^5x^2(1-6x+9x^2)+ \\ +35.2^4x^3(1-9x+\ldots)+35.2^3x^4(1-\ldots)$$

$$= 2^3[16+56(x-3x^2)+84x^2(1-6x+9x^2)+ \\ +70x^3(1-9x+\ldots)+35x^4(1-\ldots)]$$

$$= 2^3[16+56x-84x^2-434x^3+161x^4+\ldots].$$

† As the reader becomes more familiar with this type of sum he will probably teach himself to leave out many of the intermediate steps here given.

5.5. *Approximations.* Consider any particular polynomial in x, say

$$f(x) \equiv 3+4x-5x^2+6x^3-9x^4. \tag{1}$$

When x is small, the value of x^2 is smaller, and the values of x^3 and x^4 smaller still. The expression $3+4x$ differs from $f(x)$ by $x^2(-5+6x-9x^2)$, which is small compared with x. Thus $3+4x$ is an approximation to the actual value of $f(x)$ when x is small.

Similarly, $3+4x-5x^2$ is an approximation to the actual value of $f(x)$ when x is small; it differs from $f(x)$ by $x^3(6-9x)$, which is small compared with x^2.

We speak of $3+4x$ as an approximation correct to terms in x, of $3+4x-5x^2$ as an approximation correct to terms in x^2; and so on.

NOTATION. We use the symbol $\simeq$ to denote 'is approximately equal to'.

PROBLEM. *Obtain the approximation correct to terms in x^2 of $(1-x+3x^5)^7$.*

SOLUTION. Neglecting all terms of degrees above the second, we have

$$(1-x+3x^5)^7 \simeq (1-x)^7 \simeq 1-7x+21x^2,$$

the last step being derived from the binomial theorem.

6. The product of two polynomials

6.1. It is often necessary to write down some (or all) of the terms that arise when two polynomials are multiplied together. For example, to multiply

$$x^3+3ax^2+3a^2x+a^3 \quad \text{by} \quad x^4-5bx^3+b^4$$

we write down

$$\begin{aligned}(x^3+3ax^2+3a^2x+a^3)(x^4-5bx^3+b^4) \qquad (1)\\ = x^7+x^6(3a-5b)+x^5(3a^2-15ab)+x^4(a^3-15a^2b)+\\ +x^3(b^4-5a^3b)+3ab^4x^2+3a^2b^4x+a^3b^4.\end{aligned}$$

The manner of writing down this last expression is best explained by examining the coefficient of a particular power of x, say of x^6. We get an x^6

(i) when we multiply the x^3 of the first bracket of (1) by the $-5bx^3$ of the second bracket,

(ii) when we multiply the $3ax^2$ of the first bracket of (1) by the x^4 of the second bracket;

and no other x^6 occurs in the multiplication. Therefore the coefficient of x^6 in the product is $-5b+3a$.

Sometimes all we want is the coefficient of a particular power of x and then, of course, we do not write down all the terms, but only the one we want. For example, the coefficient of x^3 in the product of

$$(x^3+3x^2+3x+1)(x^4-4x^3+6x^2-4x+1)$$

is seen to be $1-3.4+3.6-4$, i.e. 3, for we get an x^3 in the product in the four ways

$$x^3\times 1, \qquad 3x\times 6x^2,$$
$$3x^2\times(-4x), \qquad 1\times(-4x^3).$$

6.2. The following type of problem serves to provide practice in manipulation of this sort.

PROBLEM. *Find the values of a and b in order that, when $(x+a)^3(x-b)^6$ is expanded in powers of x, the coefficient of x^8 shall be zero and the coefficient of x^7 shall be -9.*

SOLUTION.

$$(x+a)^3(x-b)^6 = (x^3+3ax^2+3a^2x+a^3)(x^6-6bx^5+15b^2x^4-...).$$

[*Note.* We want only x^8 and x^7, so that there is no point in continuing the second bracket beyond x^4; any further term cannot lead to x^7 or x^8 in the product.]

This is equal to

$$x^9+x^8(3a-6b)+x^7(3a^2-18ab+15b^2)+....$$

We must therefore choose a and b so that

$$3a-6b = 0, \qquad 3a^2-18ab+15b^2 = -9.$$

These equations give $a = 2b$, and $-9b^2 = -9$. Hence the required values of a and b are EITHER $b = +1$, $a = +2$; OR $b = -1$, $a = -2$.

EXAMPLES VI B

Examples on § 4

1. Find the expansions in ascending powers of x of

$$(2+x)^4, \quad (3+2x)^5, \quad (x+1)^6, \quad (2x+1)^5.$$

2. Find the expansions in descending powers of x of

$$(x+1)^7,\ (x+2)^5,\ (2x+3)^3,\ (1+x)^8.$$

3. Find the coefficient of x^7 in the expansions of

$$(2+x)^8,\ (2x+3)^8,\ (x+1)^9.$$

4. Find the coefficient of x^2 in the expansions of

$$(1+x)^{10},\ (1+2x)^7,\ (2+x)^5.$$

5. Express

$$\text{(i)}\ (1+x)^7+2(1+x)^5,$$
$$\text{(ii)}\ (1+x)^6-(1+x)^5,$$

as polynomials in x.

6. Find the coefficient of x^3 in the expansion of

$$\text{(i)}\ (1+x)^7-2(1+x)^6,$$
$$\text{(ii)}\ (1+2x)^5-(2+x)^5.$$

Examples on §§ 5, 6

7. Calculate to 4 significant figures $(1{\cdot}02)^7$ and [*harder*] $(1{\cdot}11)^9$.

8. Calculate to 5 significant figures

$$(1{\cdot}002)^8, \qquad (1{\cdot}021)^7.$$

9. Calculate to 4 significant figures

$$(0{\cdot}98)^5, \qquad (0{\cdot}992)^8.$$

10. Expand $(x+x^{-1})^7$ in descending powers of x; determine the coefficients of x^{-2} and x^{-1} in the expansion of $(x+x^{-1})^7(x-x^{-1})$.

11. Expand $(x+x^{-1})^4$ and $(x+x^{-1})^4(x^2+3+x^{-2})$ in descending powers of x.

12. Express $(2+x-3x^2)^3$ as a polynomial in x, and (*harder*) find the coefficients of x^3 and x^{14} in the expansion of $(2+x-3x^2)^7$.

13. (*Harder.*) Obtain the expansion of $(3x+1+3x^{-1})^4$ in the form $a_4(x^4+x^{-4})+a_3(x^3+x^{-3})+a_2(x^2+x^{-2})+a_1(x+x^{-1})+a_0$.

14. Find the term independent of x in the expansion of

$$(2x+1)(1+2x^{-1})^7.$$

15. Obtain the approximations correct to terms in x^2, when x is small, of

$$(1-2x-3x^2)^5, \qquad (1+4x-3x^3)^6, \qquad (3+x-x^2)^3.$$

16. Find the coefficients of x^{11} and x^{10} in the expansion of

$$(x+a)^6(x-b)^6.$$

17. Find the numerical value of a which is such that the coefficients of x^7 and x^6 in the expansion of $(x+a)^5(x-2a)^3$ are equal and different from zero.

Examples on 'the general term'

18. Find the general term in the expansion of $(3-2x)^n$.

SOLUTION. The term in x^r is

$$_nC_r\, 3^{n-r}(-2x)^r = (-1)^r \frac{n(n-1)...(n-r+1)}{r!} 3^{n-r} 2^r x^r.$$

19. Find the general terms in the expansions of

$$(4+3x)^n,\ (5-2x)^n,\ (a+2b)^n.$$

20. Find the general terms in the expansions of

$$\left(2x^2+\frac{3}{x}\right)^n, \qquad \left(2x+\frac{3}{x^2}\right)^n.$$

7.* The binomial coefficients

7.1. *The greatest coefficient.* We here answer the question 'which coefficient, or coefficients, in the expansion

$$(1+x)^n = 1+nx+...+{}_nC_r x^r+...+x^n \tag{1}$$

is the greatest?' As we have seen,

$$_nC_r = \frac{n!}{(n-r)!\,r!}, \qquad {}_nC_{r+1} = \frac{n!}{(n-r-1)!\,(r+1)!}.$$

Hence
$$\frac{_nC_{r+1}}{_nC_r} = \frac{(n-r)!}{(n-r-1)!}\cdot\frac{r!}{(r+1)!} = \frac{n-r}{r+1}. \tag{2}$$

It follows from (2) that

(*a*) when $n-r > r+1$, i.e. when $r < \frac{1}{2}(n-1)$, $_nC_{r+1} > {}_nC_r$;

(*b*) when $n-r = r+1$, i.e. when $r = \frac{1}{2}(n-1)$, $_nC_{r+1} = {}_nC_r$;

(*c*) when $n-r < r+1$, i.e. when $r > \frac{1}{2}(n-1)$, $_nC_{r+1} < {}_nC_r$.

If n is odd, $r = \frac{1}{2}(n-1)-1$ is the greatest value of r to satisfy (*a*), and so

$$1 < {}_nC_1 < ... < {}_nC_{\frac{1}{2}(n-1)-1} < {}_nC_{\frac{1}{2}(n-1)};$$

from (*b*),
$$_nC_{\frac{1}{2}(n-1)} = {}_nC_{\frac{1}{2}(n+1)},$$

and, from (*c*),
$$_nC_{\frac{1}{2}(n+1)} > {}_nC_{\frac{1}{2}(n+3)} > ... > {}_nC_n.$$

The two middle terms of (1) are 'equal greatest'.

If n is even, $r = \frac{1}{2}n-1$ is the greatest value of r to satisfy (*a*), no value of r satisfies (*b*), and so

from (*a*), $\quad 1 < {}_nC_1 < ... < {}_nC_{\frac{1}{2}n-1} < {}_nC_{\frac{1}{2}n}$;

from (*c*), $\quad {}_nC_{\frac{1}{2}n} > {}_nC_{\frac{1}{2}n+1} > {}_nC_{\frac{1}{2}n+2} > ... > {}_nC_n$.

The middle term of (1) is greatest.

7.2. The same method may be applied to the solution of problems of which the following are typical; the second problem is easier than the first, which may be omitted on a first reading.

PROBLEM 1.* *Find the numerically greatest coefficient, or coefficients, in the expansion of* $(1+3x)^n$.

SOLUTION. Let the expansion be $1+a_1x+...+a_rx^r+...$. Then

$$a_{r+1} = 3^{r+1}\frac{n!}{(n-r-1)!\,(r+1)!}, \qquad a_r = 3^r\frac{n!}{(n-r)!\,r!},$$

and
$$\frac{a_{r+1}}{a_r} = \frac{3(n-r)}{r+1}.$$

It follows that

(a) when $3n-3r > r+1$, i.e. when $r < \frac{1}{4}(3n-1)$, $a_{r+1} > a_r$;

(b) when $3n-3r = r+1$, i.e. when $r = \frac{1}{4}(3n-1)$, $a_{r+1} = a_r$;

(c) when $3n-3r < r+1$, i.e. when $r > \frac{1}{4}(3n-1)$, $a_{r+1} < a_r$.

There is an integer value of r satisfying (b) only when $3n-1$ is a multiple of 4. If $3n-1 = 4k$, where k is an integer,

$$1 < a_1 < ... < a_{k-1} < a_k \quad \text{from } (a),$$
$$a_k = a_{k+1} \quad \text{from } (b),$$
and
$$a_{k+1} > a_{k+2} > ... > a_n. \quad \text{from } (c).$$

The coefficients of x^k and x^{k+1} are equal greatest.

When $3n-1$ is not a multiple of 4, denote† by $[\frac{1}{4}(3n-1)]$ the largest integer less than $\frac{1}{4}(3n-1)$. Then, from (a),

$$1 < a_1 < ... < a_{[\frac{1}{4}(3n-1)]} < a_{[\frac{1}{4}(3n-1)]+1},$$

and, from (c), the coefficients that follow are less than $a_{[\frac{1}{4}(3n-1)]+1}$.

The greatest coefficient is a_K, where K is the integer next above $\frac{1}{4}(3n-1)$. For example, if $n = 13$, $\frac{1}{4}(3n-1) = \frac{19}{2}$ and $K = 10$.

PROBLEM 2. *Find the numerically greatest coefficient in the expansion of* $(1+5x)^{13}$.

† The notation $[x]$, for the largest integer that is less than or equal to x, is commonly used in many branches of mathematics.

SOLUTION. Let the expansion be $1+a_1x+...+a_rx^r+...$. Then

$$a_{r+1} = 5^{r+1}\frac{13!}{(12-r)!\,(r+1)!}, \qquad a_r = 5^r\frac{13!}{(13-r)!\,r!},$$

and
$$\frac{a_{r+1}}{a_r} = \frac{5(13-r)}{r+1}.$$

It follows that

(a) when $65-5r > r+1$, i.e. when $6r < 64$, $a_{r+1} > a_r$;

(b) when $65-5r = r+1$, i.e. when $6r = 64$, $a_{r+1} = a_r$;

(c) when $65-5r < r+1$, i.e. when $6r > 64$, $a_{r+1} < a_r$.

Now (a) holds when $r = 0, 1,..., 10$,

and (c) holds when $r = 11, 12, 13$.

$$\therefore\quad 1 < a_1 < a_2 < ... < a_{10} < a_{11}$$

and $a_{11} > a_{12} > a_{13}$.

Hence the greatest coefficient is

$$a_{11} = 5^{11}.\frac{13!}{2!\,11!} = 13.6.5^{11}.$$

7.3. *Identities connecting binomial coefficients.* A number of identities can be obtained by giving x special values in the results

$$(1+x)^n = 1+nx+{}_nC_2x^2+...+{}_nC_rx^r+...+x^n, \tag{1}$$
$$(1-x)^n = 1-nx+{}_nC_2x^2-...+(-1)^r{}_nC_rx^r+...+(-1)^nx^n, \tag{2}$$

and, of course, in other expansions of a like nature. The results of putting $x = 1$ in (1) and (2) are

$$1+{}_nC_1+{}_nC_2+...+{}_nC_n = 2^n, \tag{3}$$
$$1-{}_nC_1+{}_nC_2-...+(-1)^n{}_nC_n = 0; \tag{4}$$

both are useful results.

Another famous identity, often called Vandermonde's theorem, comes from the simple consideration that

$$(1+x)^{m+n} = (1+x)^m\times(1+x)^n.$$

First, let r be a given number less than m. Then

$$(1+x)^m = 1+{}_mC_1x+...+{}_mC_rx^r+...+x^m. \tag{5}$$

Also, whether r is less than, equal to, or greater than n,

$$(1+x)^n = 1+{}_nC_1x+{}_nC_2x^2+... \text{ to } (n+1) \text{ terms.} \tag{6}$$

When we multiply (5) by (6) the total coefficient of x^r in the product is the sum of

$$\begin{array}{ll} \text{coeff. of } x^r & \text{in (5)} \times \text{coeff. of } x^0 \text{ in (6)}, \\ \quad ,, \quad x^{r-1} & \text{in (5)} \times \quad ,, \quad x \text{ in (6)}, \\ \quad ,, \quad x^{r-2} & \text{in (5)} \times \quad ,, \quad x^2 \text{ in (6)}, \end{array}$$

and so on; the sum continuing until we get

$$\text{coeff. of } x^0 \text{ in (5)} \times \text{coeff. of } x^r \text{ in (6)}$$

when $r \leqslant n$, and the sum continuing until we get

$$\text{coeff. of } x^{r-n} \text{ in (5)} \times \text{coeff. of } x^n \text{ in (6)}$$

when $r > n$. The coefficient of x^r in the product of (5) and (6) is thus

$$_mC_r + {_mC_{r-1}} \cdot {_nC_1} + {_mC_{r-2}} \cdot {_nC_2} + \ldots, \tag{7}$$

the sum continuing for $r+1$ terms when $r \leqslant n$, and for $n+1$ terms when $r > n$.

But the coefficient of x^r in $(1+x)^{m+n}$ is $_{m+n}C_r$, which is therefore equal to (7).

7.4. *Vandermonde's theorem (continued).* We can write (7) in a more convenient form, one better adapted for dealing with values of r greater than m or n, if we introduce the symbol $_nC_k$, where k may have any value $0, \pm 1, \pm 2, \ldots$, and agree that $_nC_0 \equiv 1$, while $_nC_k$ shall denote zero whenever k is negative or is a positive number greater than n. We may then write the result proved in § 7.3 as

$$_{m+n}C_r = {_mC_r} \cdot {_nC_0} + {_mC_{r-1}} \cdot {_nC_1} + \ldots + {_mC_0} \cdot {_nC_r}, \tag{7 a}$$

there being $r+1$ terms on the R.H.S.; when $r > n$, some of the end coefficients will be zero. The formula (7 a) is known as VANDERMONDE'S THEOREM.

With the above conventions concerning the meaning of the symbols, (7 a) remains true when $r > m$.

The particular value $n = 1$ gives

(i) $\quad _{m+1}C_r = {_mC_r} + {_mC_{r-1}} \quad ({_1C_0} = 1,\ {_1C_1} = 1,\ {_1C_2} = 0)$,

while the particular value $n = 2$ gives

(ii) $\quad _{m+2}C_r = {_mC_r} + 2\,{_mC_{r-1}} + {_mC_{r-2}}$.

On the other hand, these last two identities are best derived by an independent consideration of the coefficient of x^r in

(i) $(1+x)^{m+1}$ and in $(1+x)(1+x)^m$,

(ii) $(1+x)^{m+2}$ and in $(1+2x+x^2)(1+x)^m$.

In fact, any particular case of the theorem is best dealt with by going back to the simple fact on which the theorem rests, namely, the coefficient of x^r is the same in $(1+x)^m.(1+x)^n$ as it is in $(1+x)^{m+n}$.

7.5. *Further identities.*† A vast number of other identities, few of them having an importance comparable with those already considered, may be derived by various special devices. We shall work out four typical examples.

PROBLEM 1. *Prove that*

$${}_nC_r - {}_nC_{r-1}\cdot{}_nC_1 + {}_nC_{r-2}\cdot{}_nC_2 - \ldots \textit{ to } (r+1) \textit{ terms}$$

is equal to 0 *when* r *is odd and is equal to* $(-1)^{\frac{1}{2}r}{}_nC_{\frac{1}{2}r}$ *when* r *is even.*

SOLUTION. The form of the expression suggests the coefficient of x^r in the product of the two expansions

$$(1-x)^n = 1-nx+\ldots+(-1)^r{}_nC_rx^r+\ldots,$$
$$(1+x)^n = 1+nx+\ldots+{}_nC_rx^r+\ldots.$$

The coefficient of x^r in their product is

$$(-1)^r[{}_nC_r - {}_nC_{r-1}\cdot{}_nC_1 + {}_nC_{r-2}\cdot{}_nC_2 - \ldots]$$

and must be equal to the coefficient of x^r in the expansion of $(1-x^2)^n$. But the latter coefficient is zero when r is odd and is $(-1)^{\frac{1}{2}r}{}_nC_{\frac{1}{2}r}$ when r is even. This gives the result required.

PROBLEM 2. *Prove that*

$$1^2+({}_nC_1)^2+({}_nC_2)^2+\ldots \textit{ to } (n+1) \textit{ terms}$$

is equal to ${}_{2n}C_n$.

SOLUTION.

$$(1+x)^n = 1+nx+\ldots+{}_nC_rx^r+\ldots+x^n. \quad (1)$$

† Most readers, though not all, will probably be well advised if they omit this section and Examples VI c on a first reading. The section will probably suit the taste of a 'mathematical specialist'.

Also,

$$(1+x)^n = x^n(1+x^{-1})^n = x^n\{1+nx^{-1}+...+{}_nC_r x^{-r}+...+x^{-n}\}. \quad (2)$$

Hence, on multiplying (1) by (2),

$$1^2+n^2+...+({}_nC_r)^2+...$$

is the coefficient of x^n in the expansion of $(1+x)^n\times(1+x)^n$, i.e. in the expansion of $(1+x)^{2n}$; and this coefficient is ${}_{2n}C_n$.

PROBLEM 3. *Prove that*

$$1+2\,{}_nC_1+3\,{}_nC_2+...+n\,{}_nC_{n-1}+(n+1) = (n+2)2^{n-1}.$$

SOLUTION. Consider

$$S(x) = 1+2\,{}_nC_1\,x+3\,{}_nC_2\,x^2+...+(n+1)x^n.$$

The form of the right-hand side shows that the multipliers 1, 2, 3,..., $n+1$ of the binomial coefficients will disappear on integration. This suggests the next step.

$$\int S(x)\,dx = x+{}_nC_1\,x^2+{}_nC_2\,x^3+...+x^{n+1}+\text{const.}$$
$$= x(1+x)^n+\text{const.}$$
$$\therefore\ S(x) = \frac{d}{dx}\{x(1+x)^n\},$$

i.e. $$S(x) = (1+x)^n+nx(1+x)^{n-1}.$$

The required result follows on putting $x = 1$.

PROBLEM 4. *Prove that*

$$1+\frac{1}{2}n+\frac{1}{3}\frac{n(n-1)}{2!}+...\ \textit{to } n+1 \textit{ terms} = \frac{1}{n+1}(2^{n+1}-1).$$

SOLUTION. Let

$$S(x) = x+\frac{1}{2}nx^2+\frac{1}{3}\frac{n(n-1)}{2!}x^3+...+\frac{1}{n+1}x^{n+1}.$$

The form of the right-hand side shows that the multipliers 1, $\frac{1}{2}$; $\frac{1}{3}$,... of the binomial coefficients will disappear on differentiation. This indicates the next step; differentiate $S(x)$. We get

$$S'(x) = 1+nx+\frac{n(n-1)}{2!}x^2+...+x^n$$
$$= (1+x)^n.$$
$$\therefore\ S(x) = \int (1+x)^n\,dx = \frac{(1+x)^{n+1}}{n+1}+A,$$

where A is a constant.

But $S(x) = 0$ when $x = 0$, so that $A = -1/(n+1)$. Hence

$$S(x) = \frac{(1+x)^{n+1}-1}{n+1},$$

and the required result follows on putting $x = 1$.

EXAMPLES VI c*

1. Find the numerical value of the greatest coefficient in the expansion of (i) $(1+x)^8$, (ii) $(1+x)^{10}$.

2. Prove that in the expansion of $(1+x)^{11}$ the coefficients of x^5 and of x^6 are equal and are greater than any other coefficient.

HINT. Use the *method* but do not quote the result of § 7.1.

3. Prove that, if $(2+3x)^7 \equiv a_0+a_1x+...+a_7x^7$, then

$$\frac{a_{r+1}}{a_r} = \frac{21-3r}{2r+2},$$

and deduce that $a_0 < a_1 < ... < a_4$; $a_4 > a_5 > a_6 > a_7$.

4. Prove that in the expansion of $(5+6x)^{11}$ the coefficient of x^6 is greater than any other coefficient.

5. Find the value of the greatest term in the expansion of $(1+3x)^7$ when $x = \frac{1}{4}$.

6. (*Harder.*) Find the value of r for which the coefficients of x^{38-3r} and x^{35-3r} in the expansion of $\left(2x^2+\frac{3}{x}\right)^{19}$ are equal.

HINT. First write down the general term; Ex. 20, VI B.

7. (*Harder.*) Prove that in the expansion of $(1+2x)^{3n+2}$ the coefficients of x^{2n+1} and of x^{2n+2} are equal and are greater than the other coefficients in the expansion.

8. Prove that $$1+{}_nC_2+{}_nC_4+...,$$
where the last term is ${}_nC_n$ $(=1)$ if n is even and is ${}_nC_{n-1}$ $(=n)$ if n is odd, is equal to 2^{n-1}.

9. Obtain identities by considering the coefficient of x^r in the expansion of each side of the identities

$$(1+x)^{n+3} \equiv (1+x)^n(1+3x+3x^2+x^3),$$
$$(1+x)^{n+4} \equiv (1+x)^n(1+4x+6x^2+4x^3+x^4).$$

10. Find the coefficient of x^{2r} in the expansions of $(1+x)^4(1-x^2)^n$ and of $(1-x)^n(1+x)^{n+4}$, and hence obtain the identity

$$(-1)^r({}_nC_r-6\,{}_nC_{r-1}+{}_nC_{r-2})$$
$$= {}_nC_0\cdot{}_{n+4}C_{2r}-{}_nC_1\cdot{}_{n+4}C_{2r-1}+...+{}_nC_{2r}\cdot{}_{n+4}C_0.$$

11. (*Harder.*) Obtain an identity by considering the coefficient of x^{2r} in the expansion of each side of the identity

$$(1+x)^m(1-x^2)^n \equiv (1-x)^n(1+x)^{n+m}.$$

NOTE ON EXX. 12–20. The answers, given at the end of the book, show one method of obtaining each result.

12. Prove that

$$_mC_r \cdot {}_{m+1}C_0 - {}_mC_{r-1} \cdot {}_{m+1}C_1 + \ldots + (-1)^r {}_mC_0 \cdot {}_{m+1}C_r$$
$$= \begin{cases} (-1)^{\frac{1}{2}r} {}_mC_{\frac{1}{2}r}, \text{ when } r \text{ is even,} \\ (-1)^{\frac{1}{2}(r+1)} {}_mC_{\frac{1}{2}(r-1)}, \text{ when } r \text{ is odd.} \end{cases}$$

13. Prove that

$$_nC_0 \cdot {}_nC_r + {}_nC_1 \cdot {}_nC_{r+1} + \ldots + {}_nC_{n-r} \cdot {}_nC_n = {}_{2n}C_{n+r}.$$

14. Prove that

$$_{m+n}C_m \cdot {}_nC_0 + {}_{m+n}C_{m+1} \cdot {}_nC_1 + \ldots + {}_{m+n}C_{m+n} \cdot {}_nC_n = {}_{2n+m}C_n.$$

15. Prove that

$$_{m+n}C_0 \cdot {}_nC_0 + {}_{m+n}C_1 \cdot {}_nC_1 + \ldots + {}_{m+n}C_n \cdot {}_nC_n = {}_{2n+m}C_n.$$

16. Prove that

$$2+3\,{}_nC_1 x + 4\,{}_nC_2 x^2 + \ldots + (n+2){}_nC_n x^n = (1+x)^{n-1}\{2+(n+2)x\}.$$

17. (*Harder.*) Prove that

$$1^2+2^2\,{}_nC_1 x + 3^2\,{}_nC_2 x^2 + \ldots + (n+1)^2\,{}_nC_n x^n = \frac{d}{dx}\{x(1+x)^n + nx^2(1+x)^{n-1}\}.$$

18. Deduce from Example 17 that

$$1^2+2^2\,{}_nC_1 + 3^2\,{}_nC_2 + \ldots + (n+1)^2{}_nC_n = 2^{n-2}(n^2+5n+4).$$

19. (*Harder.*) Prove that

$$1.2+2.3\,{}_nC_1 + \ldots + (n+1)(n+2){}_nC_n = 2^{n-2}(n^2+7n+8).$$

20. Prove that, if

$$F(x) = \frac{x^2}{1.2} + n\frac{x^3}{2.3} + \frac{n(n-1)}{2!} \cdot \frac{x^4}{3.4} + \ldots \text{ to } n+1 \text{ terms,}$$

then
$$F''(x) = (1+x)^n.$$

Deduce that

$$\frac{1}{1.2} + \frac{1}{2.3}n + \frac{1}{3.4} \cdot \frac{n(n-1)}{2!} + \ldots \text{ to } n+1 \text{ terms}$$
$$= (2^{n+2}-n-3)/\{(n+1)(n+2)\}.$$

CHAPTER VII

POLYNOMIALS IN MORE THAN ONE VARIABLE

1. Homogeneous polynomials

1.1. In the polynomial

$$3x^3+4x^2y-3xy^2-7y^3$$

each separate term is of total degree 3 in x and y; each separate term is of the form $a_r x^r y^{3-r}$, where a_r is a constant. Such a polynomial is said to be homogeneous and of degree 3 in x and y.

Similarly, a polynomial in three variables x, y, z, which is the sum of terms like

$$a_{rs} x^r y^s z^{n-r-s},$$

wherein a_{rs} denotes a constant, is said to be homogeneous and of degree n in x, y, z; each term is of total degree n in x, y, z. For example,

$$3x^4+7x^2yz-5xyz^2+10z^4$$

is homogeneous and of degree 4 in x, y, z.

1.2. THEOREM 17. *If $f(x,y,z)$ is homogeneous and of degree n in x, y, z, and k is a constant, then*

$$f(kx,ky,kz) \equiv k^n f(x,y,z).$$

PROOF. By hypothesis, $f(x,y,z)$ is the sum of a number of terms like

$$a_{rs} x^r y^s z^{n-r-s}. \tag{1}$$

In order to obtain the value of $f(kx,ky,kz)$ we write kx for x, ky for y, and kz for z in the expression for $f(x,y,z)$. Thus $f(kx,ky,kz)$ is the sum of a number of terms like

$$a_{rs}(kx)^r(ky)^s(kz)^{n-r-s},$$

i.e.

$$a_{rs} x^r y^s z^{n-r-s} k^n. \tag{2}$$

Thus each term (2) is k^n times the corresponding term (1) and so $f(kx,ky,kz)$ is k^n times $f(x,y,z)$.

COROLLARY. *The theorem is true, not only for three variables x, y, z, but for any number of variables.*

For example, if $f(x,y)$ is homogeneous and of degree 6 in x and y,

$$f(kx,ky) = k^6 f(x,y);$$

and if $f(x, y, z, t)$ is homogeneous and of degree 7 in x, y, z, t,

$$f(2x, 2y, 2z, 2t) = 2^7 f(x, y, z, t).$$

Examples VII, 2 (i)–(iii) will provide practice in applications of the theorem.

2. Symmetrical functions ; $\sum$-notation

DEFINITION. *A function of two or more variables is said to be* SYMMETRICAL *in these variables when its value is unaltered by the interchange of any two of the variables.*

For example,

$$x+y+z, \qquad yz+zx+xy$$

are symmetrical in x, y, z; but $x+y+2z$ is not symmetrical in x, y, z, since it is altered by the interchange of y and z, which changes it to $x+z+2y$.

In dealing with such functions the $\sum$ notation is useful. Suppose we are dealing with three variables a, b, c; we use $\sum a$ to denote the sum of all terms of which a is a type, that is,

$$\sum a \text{ denotes } a+b+c;$$

similarly, $\sum bc$ denotes $bc+ca+ab$;

and $\sum bc(b+c)$ denotes $bc(b+c)+ca(c+a)+ab(a+b)$.

3. Alternating functions and cyclic expressions

3.1. DEFINITION. *A function of two or more variables is said to be an* ALTERNATING FUNCTION *when the interchange of any two of the variables multiplies the value of the function by* -1.

For example, $b-c$ is an alternating function of b and c; and $(b-c)(c-a)(a-b)$ is an alternating function of a, b, c.

We shall not go into detail concerning these functions; we note the property of alternating functions and call attention to the fact that it is *not* the same as the property of symmetrical functions. When two variables are interchanged, an alternating function is changed in sign but not in magnitude, while a symmetrical function is changed neither in sign nor in magnitude.

3.2. *Cyclic expressions.* We now consider another property connected with a change of variables. We begin with the particular example

$$x^2(y-z)+y^2(z-x)+z^2(x-y).$$

Write x, y, z in order round the circumference of a circle and mark in the arrows, as shown.

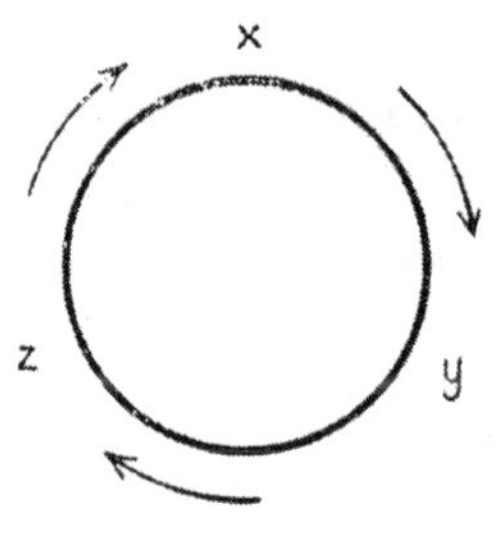

FIG. 12

First write down $x^2(y-z)$. This begins with x^2 and multiplies it by the difference of the next two letters encountered as we go round the circle. If we begin with y^2 and multiply it by the difference of the next two letters, we get $y^2(z-x)$; and if we do the same thing beginning with z^2, we get $z^2(x-y)$. The sum

$$x^2(y-z)+y^2(z-x)+z^2(x-y)$$

consists of the first term plus the two similar terms one gets by moving on round the cycle. Moreover, the sum is unaltered if throughout we write y for x, z for y, and x for z, for such a change merely alters the order in which the three terms appear.

This last property is characteristic of what is called a cyclic expression, though an expression need not be a sum of three separate terms in order to have the property. For example,

$$(y-z)(z-x)(x-y)$$

is unaltered in value if we write z for y, x for z, and y for x; such a change merely alters the order of the factors.

3.3. DEFINITION. *An expression in x, y, z which is such that its value is unaltered when we write y for x, z for y, and x for z is said to be* CYCLIC *in x, y, z.*

We refer to this change of variables as a CYCLIC CHANGE OF VARIABLES.

When we write down any cyclic expression it is advisable to observe the cyclic order of the letters; thus we write $bc+ca+ab$ and NOT $ac+ba+cb$, or again, we write abc and NOT acb.

The $\sum$ notation is again useful in dealing with cyclic expressions; when the variables are x, y, z

$$\sum x^2(y-z) \text{ denotes } x^2(y-z)+y^2(z-x)+z^2(x-y),$$
$$\sum x^2 \text{ denotes } x^2+y^2+z^2,$$

and so on.

4.* Factors and identities

4.1. The remainder theorem, coupled with a careful consideration of the degree of a polynomial in two, three, or more variables, often enables us to find the factors of the polynomial in question. We give some examples of the procedure.

PROBLEM 1. *Prove that*

$$a^2(b-c)+b^2(c-a)+c^2(a-b) \equiv -(b-c)(c-a)(a-b).$$

SOLUTION. Denote the expression on the left-hand side by F. We first think of F as a quadratic in a whose coefficients involve b and c. When we put $a = b$, $F = 0$; hence (Theorem 2) $a-b$ is a factor of F.

Similarly, we see that $b-c$ and $c-a$ are factors of F.

Since F is of total degree 3 in a, b, c, there can be no factor involving a, b, c other than the three factors $b-c$, $c-a$, $a-b$. Any other factor can involve only a numerical constant (Theorem 6, Corollary 3). Therefore

$$F \equiv k(b-c)(c-a)(a-b), \tag{1}$$

where k is a numerical constant. The value of k can be found in either of two ways, (*a*) and (*b*) below.

(*a*) The coefficient of a^2b in F is $+1$; the coefficient of a^2b in $k(b-c)(c-a)(a-b)$ is $-k$. Hence (by Theorem 6, Corollary 2) $k = -1$.

(*b*) Put $a = 0$, $b = 1$, $c = 2$ in the identity

$$a^2(b-c)+b^2(c-a)+c^2(a-b) \equiv k(b-c(c-a)(a-b).$$

We get $$0+2+4(-1) = k(-1).2.(-1),$$
i.e. $$k = -1.$$

PROBLEM 2. *Prove that*
$$\sum a^3(b-c) \equiv -(b-c)(c-a)(a-b)(a+b+c).$$

SOLUTION. As in the solution of Problem 1, we see that $(b-c)(c-a)(a-b)$ is a factor. Since the left-hand side is of degree 4 and is cyclic in a, b, c, and since the product $(b-c)(c-a)(a-b)$ is of degree 3 and is cyclic in a, b, c, it follows that the remaining factor must be of degree 1 (Theorem 6, COROLLARY 3) and must be cyclic in a, b, c. But the only expression† that is of degree 1 and cyclic in a, b, c is $k(a+b+c)$, where k is a numerical constant. Hence
$$\sum a^3(b-c) \equiv k(a+b+c)(b-c)(c-a)(a-b).$$
We find that $k = -1$ either by considering the coefficients of a^3b, or by giving a, b, c particular values (as in Problem 1).

PROBLEM 3. *Prove that*
$$(x-y-z)(y-z-x)(z-x-y)$$
$$\equiv \sum x^3 - \sum y^2z - \sum yz^2 + 2xyz. \quad (1)$$

SOLUTION. The product is homogeneous and of degree 3 in x, y, z; it is also cyclic in x, y, z. [The expansion must therefore be the sum of terms of degree 3: and the possible types of terms of degree 3 in x, y, z are

x^3, one variable taken 3 times;

y^2z, one variable twice and the next variable in the cycle once;

yz^2, one variable once and the next variable in the cycle twice;

xyz, each of the variables once.]‡

† Consider $la+mb+nc$; if it is cyclic in a, b, c, then
$$la+mb+nc \equiv lb+mc+na.$$
Hence $$(l-n)a+(m-l)b+(n-m)c \equiv 0.$$
This is an identity; it is true for *all* numerical values of a, b, c. Putting $a = 1$, $b = 0$, $c = 0$, we get $l = n$; putting $a = 0$, $b = 1$, $c = 0$, we get $l = m$. Hence the only cyclic expression of degree 1 in a, b, c is $l(a+b+c)$.

‡ Once the reader has grasped the idea, the explanation set out in [ ] may well be omitted from the solution of this and of similar problems.

Thus the expansion must be of the form

$$a(x^3+y^3+z^3)+b(y^2z+z^2x+x^2y)+c(yz^2+zx^2+xy^2)+dxyz, \quad (2)$$

where a, b, c, d are constants.

There are several methods of finding a, b, c, d. We shall give two methods; of these the first is the more straightforward, though it demands a little care.

Method 1. The term in x^3 in (1) is got by taking the term in x from each bracket; it is $x(-x)(-x) = +x^3$; and $a = 1$.

The coefficient of y^2z is $-1-1+1 = -1$ [we take y from two factors and z from the remaining factor; there are three ways of doing this]. The coefficient of yz^2 is $1-1-1 = -1$; and $b = c = -1$.

The coefficient of xyz is 2. [When we take x from the first bracket we get $2yz$ from the product of the second and third brackets; when we take x from the second or third bracket, there is no term yz in the product of the remaining two brackets.] Thus $d = 2$. This gives as the expansion of the product (1),
$$\sum x^3 - \sum y^2z - \sum yz^2 + 2xyz.$$

Method 2. Check first whether the term in y^2z is the same as that in yz^2 or different from it. If we interchange y and z in (1) we get

$$(x-z-y)(z-y-x)(y-x-z),$$

which is the same as (1). Thus (1) is symmetrical in y and z and hence $b = c$ in (2). [If b were not equal to c, the interchange of y and z would alter the value of (2).]

Having checked that $b = c$, we may write

$$(x-y-z)(y-z-x)(z-x-y)$$
$$\equiv a(x^3+y^3+z^3)+b(y^2z+z^2x+x^2y+yz^2+zx^2+xy^2)+dxyz, \quad (3)$$

where a, b, d are constants.

Put $x = 1$, $y = 0$, $z = 0$; we get $1 = a$.

Put $x = 0$, $y = 1$, $z = 1$; we get $0 = 2a+2b$, or $b = -1$.

Put $x = y = z = 1$; we get $-1 = 3a+6b+d$, or $d = 2$.

EXAMPLES VII

1. Which of the polynomials

(i) $2x^3+4x^2y+xy^2$, (ii) $3x^3-4y^4$,
(iii) $4x^2+2x+5y$, (iv) $4x^2+2xy+5y^2$,

is homogeneous in x and y?
Which of the polynomials

(v) $3x^3+4y^3+5y^2z$, (vi) $3x^3+4y+5yz$

is homogeneous in x, y, z?

2. Prove the following results by actual substitution, and without quoting Theorem 17:

(i) When $f(x,y) \equiv x^3+3x^2y+y^3$, $f(ax,ay) \equiv a^3f(x,y)$;
(ii) When $f(x,y,z) \equiv x^2+y^2-2yz$, $f(bx,by,bz) \equiv b^2f(x,y,z)$;
(iii) When $f(x,y,z,t) \equiv x^3+y^3+t^3-y^2z$, $f(2x,2y,2z,2t) \equiv 8f(x,y,z,t)$.

3. Of the following polynomials some are symmetrical in x, y, z, and some are not. State which are the symmetrical ones.

(i) $x^2+y^2+z^2$, (ii) $yz+zx+2xy$,
(iii) x^3y+y^3x+xyz, (iv) $x^3+y^3+z^3-3xyz$,
(v) $y^3z+z^3x+x^3y+yz^3+zx^3+xy^3$.

4. Prove that $bc(b-c)$ is an alternating function of b and c, and that $bc(b-c)+ca(c-a)+ab(a-b)$ is an alternating function of a, b, c.

HINT. When we interchange b and c in the latter function, we get

$$cb(c-b)+ba(b-a)+ac(a-c).$$

5. Write down the cyclic expressions in a, b, c of which the typical terms are

(i) $bc(b-c)$, (ii) $a^2(b-c)$, (iii) b^3-c^3,
(iv) $bc(b^2+c^2)$, (v) a^2b, (vi) $a(b-c)$.

6. Write down the cyclic expressions in x, y, z of which the typical terms are

(i) x^4, (ii) xy^2, (iii) $yz(y-z)$, (iv) y^2+z^2.

7. Prove that, with the $\sum$ notation of § 3.3,

$$\sum (y-z) \equiv 0, \qquad \sum x(y-z) \equiv 0,$$
$$\sum yz(z-x) \equiv \sum yz^2-3xyz.$$

8.* Prove that

$$\sum yz(y-z) \equiv -(y-z)(z-x)(x-y).$$

9.* Prove that

(i) $\sum bc(b-c) \equiv -(b-c)(c-a)(a-b) \equiv \sum a^2(b-c)$,
(ii) $\sum (b+c)(b^2-c^2) \equiv \sum bc(b-c) \equiv -(b-c)(c-a)(a-b)$.

10.* Prove that

(i) $\sum bc(b^2-c^2) \equiv -(a+b+c)(b-c)(c-a)(a-b)$,

(ii) $\sum b^2c(b-c)$ does not contain $b-c$ as a factor.

11.* Prove that, for every positive integer $n \geqslant 2$,

$$a^n(b-c)+b^n(c-a)+c^n(a-b)$$

contains $(b-c)(c-a)(a-b)$ as a factor.

12.* Prove that

$$a^n(b+c)+b^n(c+a)+c^n(a+b)$$

does not contain $b+c$ as a factor when neither a nor c is zero.

CHAPTER VIII

RATIONAL FUNCTIONS: GENERAL THEORY

1. Preliminary

1.1. A rational function is the ratio of two polynomials; in the language of elementary work, it is a *fraction in which the numerator and the denominator are polynomials.* The denominator must not be zero, for arithmetic does not include any definition of 'division by zero'. Values of x that would make the denominator zero are excluded from the theory of rational functions; for example, when we write

$$\frac{2}{2x-2} \equiv \frac{1}{x-1},$$

we shall mean that the two expressions are equal for all values of x other than 1. When $x = 1$ neither expression has a meaning.

1.2. The reader is familiar with examples of the type 'Express

$$\frac{1}{x+1} - \frac{2}{x+2}$$

as a single fraction', the working of which is

$$\frac{1}{x+1} - \frac{2}{x+2} = \frac{x+2-2(x+1)}{(x+1)(x+2)} = -\frac{x}{(x+1)(x+2)}. \quad (1)$$

It is always possible to express the sum (or difference) of a number of fractions as one single fraction.

In many branches of mathematics we need to be able to perform the reverse operation, namely, 'given a single fraction whose denominator breaks up into factors, to express the fraction as the sum or difference of a number of simpler fractions'.

This reverse operation is called splitting the given fraction into *Partial Fractions*, that is, finding the fractions that form the separate parts of the given fraction.

1.3. Before trying to find these partial fractions, we first look carefully at the kind of result we get when we add and subtract certain types of fractions. The usual methods of elementary algebra give†

$$\frac{1}{x+1}-\frac{1}{x+2}\equiv\frac{x+2-(x+1)}{(x+1)(x+2)}\equiv\frac{1}{(x+1)(x+2)},$$

$$\frac{3}{2x+5}+\frac{2}{7x-4}\equiv\frac{3(7x-4)+2(2x+5)}{(2x+5)(7x-4)}\equiv\frac{25x-2}{(2x+5)(7x-4)}.$$

On the left-hand side the degree of each denominator is unity, while each numerator is a constant; that is, in each fraction the degree of the numerator is less than the degree of the denominator. In the fraction on the right-hand side the degree of the numerator is again less than the degree of the denominator.

Again, elementary calculation gives

$$\frac{x+1}{x^2+1}+\frac{2x}{x^2+2}\equiv\frac{x^3+x^2+2x+2+2x^3+2x}{(x^2+1)(x^2+2)}=\frac{3x^3+\ldots+2}{(x^2+1)(x^2+2)}.$$

Here too, in the resulting fraction on the right, the degree of the numerator is less than the degree of the denominator; and it is the sum of fractions having the same property, the degree of the numerator less than the degree of the denominator.

When the degree of the numerator is less than that of the denominator we refer to the fraction as a 'PROPER FRACTION'. This technical term will be used for the sake of brevity.

1.4. The examples of § 1.3 indicate that *any given 'proper' fraction whose denominator factorizes can be written as the sum (or difference) of two or more 'proper' partial fractions.* This indicated result can be proved, but its proof involves the use of theorems that are too advanced for inclusion in this book. All we do here is to note the fact that, when we start from a proper fraction whose denominator factorizes, we can express that fraction as the sum of two or more proper partial fractions. We use this fact in all our examples.

2. Method of finding partial fractions

We shall work numerical examples and make no attempt (at this stage) to deal with general theory.

† We use the identity sign to denote the fact that the equality holds independently of the value of x.

NOTE. In the text of the following examples the words within square brackets are by way of explanation to the reader; they do not form a necessary part of the solution when the reader is working out examples for himself.

2.1. *Denominator the product of linear factors.*

PROBLEM 1. *Express*

$$\frac{2x+3}{(x-1)(x-2)(2x-3)}$$

as the sum of partial fractions.

SOLUTION. Let

$$\frac{2x+3}{(x-1)(x-2)(2x-3)} \equiv \frac{A}{x-1}+\frac{B}{x-2}+\frac{C}{2x-3}, \tag{1}$$

where A, B, C are constants. [*Explanation.* The fraction on the left is 'proper'. The degree of the denominator of each fraction on the right-hand side is *one*; to make these fractions 'proper' each numerator must be of degree less than one, and so must be a constant.]

Multiply (1) throughout by $(x-1)(x-2)(2x-3)$; we get

$$2x+3 \equiv A(x-2)(2x-3)+B(x-1)(2x-3)+C(x-1)(x-2).$$

[*Explanation.* We now substitute, in turn, the values of x that make the factors $x-1$, $x-2$, $2x-3$ equal zero. *Note these substitutions.*]

On substituting in this identity

$x = 1$, we get $5 = A(-1)(-1)$, i.e. $A = 5$;
$x = 2$, we get $7 = B(1)(1)$, i.e. $B = 7$;
$x = \frac{3}{2}$, we get $6 = C(\frac{1}{2})(-\frac{1}{2})$, i.e. $C = -24$.

Hence

$$\frac{2x+3}{(x-1)(x-2)(2x-3)} \equiv \frac{5}{x-1}+\frac{7}{x-2}-\frac{24}{2x-3}.$$

2.2. *Denominator having a quadratic factor.*

PROBLEM 2. *Express*

$$\frac{3x^2+1}{(x+1)(x-1)(x^2+1)}$$

as the sum of partial fractions.

SOLUTION. Let

$$\frac{3x^2+1}{(x+1)(x-1)(x^2+1)} \equiv \frac{A}{x+1}+\frac{B}{x-1}+\frac{Cx+D}{x^2+1}, \tag{1}$$

where A, B, C, D are constants. [*Explanation.* The given fraction is 'proper', and so we make the degree of each numerator on the right less than the degree of the corresponding denominator.]

Multiply (1) throughout by $(x+1)(x-1)(x^2+1)$; we get

$$3x^2+1 \equiv A(x-1)(x^2+1)+B(x+1)(x^2+1)+ \\ +(Cx+D)(x+1)(x-1). \tag{2}$$

[*Explanation.* We first substitute the values of x that make the *linear* factors $x+1$, $x-1$ equal to zero.]

Substituting in this identity

$x = -1$, we get $4 = A(-2)(2)$, i.e. $A = -1$;

$x = 1$, we get $4 = B(2)(2)$, i.e. $B = 1$.

[This leaves C and D still to be found. Since (2) is an identity, the coefficient of each power of x on the one side is equal to the coefficient of that same power of x on the other side (Theorem 5, Corollary 2). This fact enables us to find C and D.]

Since (2) is an identity we have, by equating the coefficients of x^3 in (2),

$$0 = A+B+C, \tag{3}$$

constant terms in (2),

$$1 = -A+B-D. \tag{4}$$

But $A = -1$ and $B = 1$; hence (3) and (4) give

$$C = 0, \qquad D = 1.$$

Hence

$$\frac{3x^2+1}{(x+1)(x-1)(x^2+1)} \equiv -\frac{1}{x+1}+\frac{1}{x-1}+\frac{1}{x^2+1}.$$

[*Explanation.* Instead of equating the coefficients of x^3 and the constant terms in (2), we could equate coefficients of x^2 and coefficients of x; but we select the easiest pair. We need only *two* of the four equations that can be got by equating coefficients in (2) because we know A and B already and we are trying to find the values of the *two* constants C and D.]

PROBLEM 3. *Express*

$$\frac{2x^2+x-1}{(x-1)(x^2+x+1)}$$

as the sum of partial fractions.

SOLUTION. [We omit explanations; the principles are the same as in Problems 1 and 2.]

Let $$\frac{2x^2+x-1}{(x-1)(x^2+x+1)} \equiv \frac{A}{x-1}+\frac{Bx+C}{x^2+x+1},$$

where A, B, C are constants. Then

$$2x^2+x-1 \equiv A(x^2+x+1)+(Bx+C)(x-1). \tag{1}$$

Put $x = 1$; we get $2 = 3A$, $A = \frac{2}{3}$.

Equate in (1)

coeffs. of x^2; we get $2 = A+B$, i.e. $B = \frac{4}{3}$;

const. terms; we get $-1 = A-C$, i.e. $C = \frac{5}{3}$.

Hence $$\frac{2x^2+x-1}{(x-1)(x^2+x+1)} \equiv \frac{2}{3(x-1)}+\frac{4x+5}{3(x^2+x+1)}.$$

2.3. *Fractions having a square (cube) in the denominator.*

PROBLEM 4. *Express*

$$\frac{3x^2+2x-1}{(x-1)^2(x-2)} \tag{1}$$

as the sum of partial fractions.

Explanation. In this problem one factor of the denominator is squared. Before we work out the solution, we explain the reason for the form of partial fraction used in the solution. Since the given fraction is 'proper', the partial fractions must be 'proper'; that is, the partial fractions are of the form

$$\frac{A}{x-2}+\frac{Bx+C}{(x-1)^2}, \tag{2}$$

where A, B, C are constants. We can, however, break down the last fraction into two simpler fractions. We have, in fact,

$$\frac{Bx+C}{(x-1)^2} = \frac{B(x-1)+B+C}{(x-1)^2} = \frac{B}{x-1}+\frac{B+C}{(x-1)^2}.$$

or, on writing D instead of $B+C$, we see that (1) is of the form

$$\frac{A}{x-2}+\frac{B}{x-1}+\frac{D}{(x-1)^2},$$

where A, B, D are constants. THIS IS THE FORM APPROPRIATE TO A FRACTION LIKE (1). When the denominator contains a cubed factor (Problem 5) the procedure is similar.

SOLUTION.

Let $$\frac{3x^2+2x-1}{(x-1)^2(x-2)} \equiv \frac{A}{x-2}+\frac{B}{x-1}+\frac{D}{(x-1)^2},$$

where A, B, D are constants. Then

$$3x^2+2x-1 \equiv A(x-1)^2+B(x-2)(x-1)+D(x-2).$$

Put $x = 2$; we get $\quad 15 = A$;

$x = 1$; we get $\quad 4 = -D$.

Equate coefficients of x^2; we get $3 = A+B$, so that

$$B = 3-15 = -12.$$

Hence $$\frac{3x^2+2x-1}{(x-1)^2(x-2)} \equiv \frac{15}{x-2}-\frac{12}{x-1}-\frac{4}{(x-1)^2}.$$

PROBLEM 5. (HARDER.)* *Express*

$$\frac{4x^3+1}{(x-1)^3(2x-1)(x-2)}$$

as the sum of partial fractions.

SOLUTION. Let

$$\frac{4x^3+1}{(x-1)^3(2x-1)(x-2)} \equiv \frac{A}{x-1}+\frac{B}{(x-1)^2}+\frac{C}{(x-1)^3}+\frac{D}{2x-1}+\frac{E}{x-2},$$

where A, B,..., E are constants. Then

$$4x^3+1 \equiv A(x-1)^2(2x-1)(x-2)+B(x-1)(2x-1)(x-2)$$
$$+C(2x-1)(x-2)+D(x-1)^3(x-2)+E(x-1)^3(2x-1).$$

Put

$x = 1$; we get $\quad 5 = C.1.(-1)$, $\quad$ i.e. $C = -5$;

$x = \frac{1}{2}$; we get $\quad \frac{3}{2} = D.(-\frac{1}{8})(-\frac{3}{2})$, $\quad$ i.e. $D = 8$;

$x = 2$; we get $\quad 33 = E.3$, $\quad$ i.e. $E = 11$.

Equate

coeffs. of x^4; we get

$$0 = 2A+D+2E, \quad \text{i.e. } A = -15;$$

const. terms; we get

$$1 = 2A-2B+2C+2D+E,$$

i.e.
$$1 = -30-2B-10+16+11,$$

or
$$B = -7.$$

Hence

$$\frac{4x^3+1}{(x-1)^3(2x-1)(x-2)}$$

$$\equiv -\frac{15}{x-1}-\frac{7}{(x-1)^2}-\frac{5}{(x-1)^3}+\frac{8}{2x-1}+\frac{11}{x-2}.$$

2.4. *Method of checking the algebra.* The simplest check on one's work is to test the answer for some particular value of x; usually $x = 0$ will serve, though it will not always be an adequate check if we have used the step 'equate the constant terms' in the course of the work [for this step is equivalent to putting $x = 0$ in the identity]. For example,

Check, Problem 1. When $x = 0$,

$$\text{L.H.S} = \frac{3}{-6} = -\frac{1}{2};$$

$$\text{R.H.S.} = -5-\frac{7}{2}+8 = -\frac{1}{2}.$$

Check, Problem 2. When $x = 2$,

$$\text{L.H.S.} = \frac{13}{3.1.5} = \frac{13}{15};$$

$$\text{R.H.S.} = -\frac{1}{3}+1+\frac{1}{5} = \frac{-5+15+3}{15} = \frac{13}{15}.$$

3. Improper fractions

3.1. When the degree of the numerator is equal to or greater than the degree of the denominator, we say that the fraction is 'improper'. An example is

$$\frac{3x^2+5x+7}{(x-1)(x-2)}.$$

When we divide $3x^2+5x+7$ by $(x-1)(x-2)$, i.e. by x^2-3x+2, the quotient is 3 and the remainder is of degree *one* in x. In fact,

$$3x^2+5x+7 \equiv 3(x^2-3x+2)+14x+1,$$

and
$$\frac{3x^2+5x+7}{(x-1)(x-2)} \equiv 3+\frac{14x+1}{(x-1)(x-2)}. \tag{1}$$

By § 2 we know that we can write (1) in the form

$$\frac{3x^2+5x+7}{(x-1)(x-2)} \equiv 3+\frac{A}{x-1}+\frac{B}{x-2}, \tag{2}$$

where A and B are constants. *We do not need to work out the remainder* '$14x+1$' *in order to find A and B*; the following problem gives the method of procedure.

PROBLEM 6. *Express*

$$\frac{3x^2+5x+7}{(x-1)(x-2)} \text{ in the form } C+\frac{A}{x-1}+\frac{B}{x-2},$$

where C, A, B are constants.

SOLUTION. C is the quotient when $3x^2+5x+7$ is divided by x^2-3x+2; this quotient is 3 and so $C = 3$. Hence

$$\frac{3x^2+5x+7}{(x-1)(x-2)} \equiv 3+\frac{A}{x-1}+\frac{B}{x-2},$$

where A and B are constants, and

$$3x^2+5x+7 \equiv 3(x-1)(x-2)+A(x-2)+B(x-1).$$

Put
$$x = 1;\text{ we get } \quad 15 = -A;$$
$$x = 2;\text{ we get } \quad 29 = B.$$

Hence
$$\frac{3x^2+5x+7}{(x-1)(x-2)} \equiv 3-\frac{15}{x-1}+\frac{29}{x-2}.$$

3.2. It may be worth while to prove that we cannot find constants A and B such that

$$\frac{3x^2+5x+7}{(x-1)(x-2)} \equiv \frac{A}{x-1}+\frac{B}{x-2}.$$

To prove this we observe that the sum of the two fractions on the right-hand side is

$$\frac{A(x-2)+B(x-1)}{(x-1)(x-2)} \equiv \frac{(A+B)x-(2A+B)}{(x-1)(x-2)}.$$

But $3x^2+5x+7$ is of degree 2, while $(A+B)x-(2A+B)$ is of degree 1; hence the two cannot be identically equal [Theorem 6, Corollary 3].

This example is sufficient to show that when the degree of the numerator is not less than the degree of the denominator, the form of partial fraction used in § 2 is not adequate by itself; the original fraction is then the sum of two parts, (1) the quotient of the numerator by the denominator, (2) the partial fractions corresponding to the factors of the denominator.

Examples VIII a

1. Find the numerical values of the constants A, B, C, D in the following identities:

(i) $\dfrac{3x^2+4}{(x-1)(x-2)(x-3)} \equiv \dfrac{A}{x-1}+\dfrac{B}{x-2}+\dfrac{C}{x-3}$,

(ii) $\dfrac{2x^2-3}{(x-1)(x-2)^2} \equiv \dfrac{A}{x-1}+\dfrac{B}{x-2}+\dfrac{C}{(x-2)^2}$,

(iii) $\dfrac{x^3+1}{(x-1)(2x-1)(x^2+1)} \equiv \dfrac{A}{x-1}+\dfrac{B}{2x-1}+\dfrac{Cx+D}{x^2+1}$.

Hint. Note the substitutions used in the examples of § 2, followed by 'equating coefficients' when necessary.

2. Express as sums of partial fractions:

(i) $\dfrac{5x^2-5x+2}{(x-1)(x+1)(2x-1)}$, (ii) $\dfrac{2}{(x-1)(x-2)(x-3)}$,

(iii) $\dfrac{2}{(x+1)(x+2)(x+3)}$, (iv) $\dfrac{2}{(x-3)(x-4)(x-5)}$.

Hint. Use the method of Problem 1.

3. (*Harder.*) Express as sums of partial fractions:

(i) $\dfrac{3x^2-2}{(x-1)(2x-1)(x+3)}$, (ii) $\dfrac{4x+5}{(x-1)(3x+2)(x-3)}$,

(iii) $\dfrac{3x-1}{(x-2)(x+2)(2x-5)}$, (iv) $\dfrac{3x^2+4}{(x-2)(x-3)(x-4)}$.

4. Express as sums of partial fractions:

(i) $\dfrac{x+3}{(x-1)(x^2+1)}$, (ii) $\dfrac{2x^2+3}{(x-2)(x^2+x+5)}$,

(iii) $\dfrac{3x^2-2}{(2x-1)(4x^2+5)}$, (iv) $\dfrac{x^3+x^2+3}{(x-1)(x-2)(x^2+1)}$.

Hint. Use the method of Problems 2 and 3.

5. By using the method of Problems 4 and 5 find the numerical values of the constants A, B, C, D in the following identities:

(i) $$\frac{1}{(x-1)^2(x-2)} \equiv \frac{A}{x-1}+\frac{B}{(x-1)^2}+\frac{C}{x-2},$$

(ii) $$\frac{x}{(x-1)(x-2)^2} \equiv \frac{A}{x-1}+\frac{B}{x-2}+\frac{C}{(x-2)^2},$$

(iii) $$\frac{1}{(x-1)^3(x-2)} \equiv \frac{A}{x-1}+\frac{B}{(x-1)^2}+\frac{C}{(x-1)^3}+\frac{D}{x-2},$$

(iv) $$\frac{x+1}{x^2(x-1)^2} \equiv \frac{A}{x}+\frac{B}{x^2}+\frac{C}{x-1}+\frac{D}{(x-1)^2}.$$

NOTE. In the concluding stages of (iv) equate coefficients of x^3 and of x^2. Equating constant terms will give the same result as putting $x = 0$, which comes in the earlier stages of the sum.

6. Express as the sum of partial fractions:

(i) $\dfrac{1}{(x-2)^2(x-3)}$, (ii) $\dfrac{x+1}{x(x-1)^2}$,

(iii) $\dfrac{1}{(x-2)^3(x-3)}$, (iv) $\dfrac{x}{(x-1)^2(x-2)^2}$.

HINT. Use the method of Problems 4 and 5; (ii) is like 5 (iv), in which 'equating constant terms' at the end gives no new information.

7. *Various types.* Express as the sum of partial fractions:

(i) $\dfrac{9x^2-52x+72}{(x-2)(x-3)(x-4)}$, (ii) $\dfrac{4-x}{x(x-1)(x-2)}$,

(iii) $\dfrac{3x^2+x+2}{(x-1)^2(x+1)}$, (iv) $\dfrac{x+2}{(2x-1)(x^2+1)}$,

(v) $\dfrac{x-1}{(x-2)(2x^2+1)}$, (vi) $\dfrac{5x^2-6x+2}{(x-1)^3(2x-1)}$.

8. Given that
$$\frac{5-x}{(x-1)(x-2)(x-3)} \equiv \frac{2}{x-1}-\frac{3}{x-2}+\frac{1}{x-3},$$
write down, by inspection, the partial fractions of

(i) $\dfrac{4-x}{x(x-1)(x-2)}$, (ii) $\dfrac{6-x}{(x-2)(x-3)(x-4)}$.

9. (*Easy*, but of frequent use.) Express as the sum of partial fractions:

(i) $\dfrac{1}{x^2-1}$, (ii) $\dfrac{2x}{x^2-1}$.

10. Express each of the fractions

$$\frac{x^2+2x-5}{(x-1)(x-2)}, \qquad \frac{x^2-2x+3}{(x-1)(x-2)}$$

in the form

$$A+\frac{B}{x-1}+\frac{C}{x-2},$$

where A, B, C are constants.

11. *Inventing your own examples.* In one or more of Questions 2, 3, 6, 7

first replace x by $x-1$;

then express the result as a sum of partial fractions;

then check your answer by replacing x by $x-1$ in the answers to the original question.

12. *Inventing your own examples.* Write down a simple sum of fractions, such as

$$\frac{1}{x-2}+\frac{2}{(x-3)^2}+\frac{3}{(x-3)^3},$$

and express the sum as a single fraction; put this fraction into partial fractions, when you should obtain the fractions you started from.

4.* Equal ratios

4.1. In many parts of mathematics, notably in algebra and in algebraic (or coordinate) geometry, it is useful to know how to manipulate two or more equal fractions. The theorems that follow provide the key to this manipulation. The student should be watchful for opportunities to use the theorems in order to simplify his algebraic work.

4.2. THEOREM 18. *If* $\dfrac{N_1}{D_1} = \dfrac{N_2}{D_2}$, *and* a, b, c, d *are any constants, then*

$$\frac{aN_1+bD_1}{cN_1+dD_1} = \frac{aN_2+bD_2}{cN_2+dD_2}.$$

PROOF. Let

$$\frac{N_1}{D_1} = \frac{N_2}{D_2} = t.$$

Then $N_1 = D_1 t$ and $N_2 = D_2 t$. Hence

$$\text{(i)} \quad \frac{aN_1+bD_1}{cN_1+dD_1} = \frac{D_1(at+b)}{D_1(ct+d)} = \frac{at+b}{ct+d},$$

$$\text{(ii)} \quad \frac{aN_2+bD_2}{cN_2+dD_2} = \frac{D_2(at+b)}{D_2(ct+d)} = \frac{at+b}{ct+d},$$

and so
$$\frac{aN_1+bD_1}{cN_1+dD_1}=\frac{aN_2+bD_2}{cN_2+dD_2}.$$

NOTE. In view of our remarks in § 1.1, we may divide by D_1 and D_2, which are not equal to zero.

4.3. *Some examples.*

(i) Let
$$\frac{x+1+\sqrt{(x^2+1)}}{x+1-\sqrt{(x^2+1)}}=\frac{y-z}{y+z}.$$

We can simplify the shape of this by applying Theorem 18 with $a=1$, $b=1$ (i.e. adding numerator and denominator to form a new numerator) and $c=1$, $d=-1$ (i.e. subtracting denominator from numerator to form a new denominator). This gives
$$\frac{2(x+1)}{2\sqrt{(x^2+1)}}=\frac{2y}{-2z},$$
or
$$\frac{x+1}{\sqrt{(x^2+1)}}=-\frac{y}{z}.$$

(ii) Let
$$\frac{y}{1}=\frac{\sqrt{(1+x)}+\sqrt{x}}{\sqrt{(1+x)}-\sqrt{x}}.$$

Then
$$\frac{y+1}{y-1}=\frac{\{\sqrt{(1+x)}+\sqrt{x}\}+\{\sqrt{(1+x)}-\sqrt{x}\}}{\{\sqrt{(1+x)}+\sqrt{x}\}-\{\sqrt{(1+x)}-\sqrt{x}\}}$$
$$=\frac{\sqrt{(1+x)}}{\sqrt{x}}.$$

(iii) Let
$$\frac{\alpha+2\beta}{2\alpha+\beta}=\frac{x+2y}{2x+y}.$$

We notice that
$$\alpha+2\beta-2(2\alpha+\beta)=-3\alpha \quad \text{(containing } \alpha \text{ only)},$$
$$2\alpha+\beta-2(\alpha+2\beta)=-3\beta \quad \text{(containing } \beta \text{ only)}.$$

We therefore apply Theorem 18 in the form
$$\frac{N_1-2D_1}{-2N_1+D_1}=\frac{N_2-2D_2}{-2N_2+D_2}.$$

This gives
$$\frac{\alpha}{\beta}=\frac{x}{y}.$$

4.4. THEOREM 19. *If there are r equal fractions,*

$$\frac{N_1}{D_1} = \frac{N_2}{D_2} = \dots = \frac{N_r}{D_r}$$

and a, b,..., k are any r numbers, then each fraction is also equal to

$$\frac{aN_1+bN_2+\dots+kN_r}{aD_1+bD_2+\dots+kD_r}$$

provided this denominator is not zero.

PROOF. Let

$$\frac{N_1}{D_1} = \frac{N_2}{D_2} = \dots = \frac{N_r}{D_r} = t.$$

Then $N_1 = D_1 t$, $N_2 = D_2 t$,..., $N_r = D_r t$, and

$$\frac{aN_1+bN_2+\dots+kN_r}{aD_1+bD_2+\dots+kD_r} = \frac{t(aD_1+bD_2+\dots+kD_r)}{aD_1+bD_2+\dots+kD_r} = t,$$

which proves the theorem.†

4.5. The next theorem covers the one case excluded by the theorem we have just proved.

THEOREM 20. *If there are r equal fractions*

$$\frac{N_1}{D_1} = \frac{N_2}{D_2} = \dots = \frac{N_r}{D_r} \quad (N_1 \neq 0),$$

and the r numbers a, b,..., k are chosen so that

$$aN_1+bN_2+\dots+kN_r = 0,$$

then also $$aD_1+bD_2+\dots+kD_r = 0.$$

COROLLARY. *If the r numbers a, b,..., k are chosen so that*

$$aD_1+bD_2+\dots+kD_r = 0,$$

then also $$aN_1+bN_2+\dots+kN_r = 0.$$

PROOF. Let

$$\frac{N_1}{D_1} = \frac{N_2}{D_2} = \dots = \frac{N_r}{D_r} = t.$$

Then $N_1 = D_1 t$, $N_2 = D_2 t$,..., $N_r = D_r t$. Moreover, since $N_1 \neq 0$, it follows that $t \neq 0$. Now, the numbers $a, b,..., k$ having been chosen so that

$$aN_1+bN_2+\dots+kN_r = 0,$$

† Notice that the last step uses division by $aD_1+bD_2+\dots+kD_r$; if this were zero, the step would not be valid.

we have $$t(aD_1+bD_2+...+kD_r) = 0.$$

But $t \neq 0$; $$\therefore\ aD_1+bD_2+...+kD_r = 0.$$

PROOF OF COROLLARY. The numbers $a, b, ..., k$ having been chosen so that

$$aD_1+bD_2+...+kD_r = 0,$$

we have $$t(aD_1+bD_2+...+kD_r) = 0,$$

i.e. $$aN_1+bN_2+...+kN_r = 0.$$

4.6. *Some examples.*

(i) PROBLEM. *Find $x:y:z$ given that*

$$\frac{x-y+z}{3} = \frac{y-z+x}{4} = \frac{z-x+y}{5}.$$

SOLUTION. Each fraction is equal to

$$\frac{N_1+N_2+N_3}{D_1+D_2+D_3}.$$

Thus $$\frac{x-y+z}{3} = \frac{x+y+z}{12}.$$

On applying Theorem 19 to these two equal fractions,

$$\frac{x+y+z}{12} = \frac{(x+y+z)-(x-y+z)}{9} = \frac{2y}{9}.$$

Working similarly with the other two fractions we get

$$\frac{2y}{9} = \frac{2z}{8} = \frac{2x}{7}.$$

Hence $$x:y:z = 7:9:8.$$

NOTE. The method is not *essential* to the solution of this sum: it is neater and quicker than the more laborious method of solving

$$4(x-y+z) = 3(y-z+x),$$
$$4(z-x+y) = 5(y-z+x).$$

(ii) PROBLEM. (HARDER.) *Find values of x, y, z to satisfy the equations*

$$\frac{x^2-yz}{1} = \frac{y^2-zx}{2} = \frac{z^2-xy}{3} = 9.$$

SOLUTION. Since each fraction is equal to $\dfrac{N_1-N_2}{D_1-D_2}$ and also to $\dfrac{N_2-N_3}{D_2-D_3}$,

$$\frac{x^2-yz-y^2+zx}{-1} = \frac{y^2-zx-z^2+xy}{-1},$$

or
$$(x-y)(x+y+z) = (y-z)(x+y+z).$$

Hence, either

$$(a)\quad x+y+z = 0, \quad \text{when } y = -(x+z),$$

or
$$(b)\quad x-2y+z = 0, \quad \text{when } y = \tfrac{1}{2}(x+z).$$

(a) When $y = -(x+z)$,

$$2(x^2-yz) = y^2-zx$$

gives, on substituting for y and simplifying,

$$x^2+zx+z^2 = 0,$$

i.e.
$$(x-z\omega)(x-z\omega^2) = 0,$$

where ω is a complex cube root of unity (p. 39). Hence,

either
$$x = z\omega \text{ and } y = -z(1+\omega) = z\omega^2,$$

or
$$x = z\omega^2 \text{ and } y = -z(1+\omega^2) = z\omega.$$

In the first case the given equations are satisfied only if $(z\omega)^2-\omega^2 z.z = 9$. Hence the solution

$$x:y:z = \omega:\omega^2:1,$$

though it satisfies $x^2-yz = (y^2-zx)/2 = (z^2-xy)/3$, will not satisfy the given equations as a whole.

Similarly, $x = z\omega^2$, $y = z\omega$ does not satisfy the given equations.

(b) When $y = \frac{1}{2}(x+z)$,

$$2(x^2-yz) = y^2-zx$$

gives, on substituting for y and simplifying,

$$7x^2-2zx-5z^2 = 0;$$

i.e.
$$(7x+5z)(x-z) = 0.$$

Hence, either $x = z$, $y = z$, which is not a solution of the given equations as a whole, or

$$x = -\tfrac{5}{7}z, \qquad y = \tfrac{1}{7}z.$$

The latter is equivalent to

$$x = -5t, \qquad y = t, \qquad z = 7t,$$

where t is an arbitrary number. This satisfies the equations if

$$\frac{(25-7)t^2}{1} = \frac{(1+35)t^2}{2} = \frac{(49+5)t^2}{3} = 9;$$

i.e. if

$$t^2 = \tfrac{1}{2}.$$

Hence the solutions of the equations are

$$x = -\frac{5}{\sqrt{2}}, \qquad y = \frac{1}{\sqrt{2}}, \qquad z = \frac{7}{\sqrt{2}}$$

and

$$x = \frac{5}{\sqrt{2}}, \qquad y = -\frac{1}{\sqrt{2}}, \qquad z = -\frac{7}{\sqrt{2}}.$$

(iii) PROBLEM. *Find the solutions of the equations*

$$\frac{x+y}{2} = \frac{y+z}{7} = \frac{z+x}{5}.$$

SOLUTION. Since $7-5-2 = 0$, we must have (Theorem 20)

$$y+z-(z+x)-(x+y) = 0,$$

i.e.

$$x = 0.$$

This gives

$$\frac{y}{2} = \frac{y+z}{7} = \frac{z}{5}$$

and these are satisfied if $5y = 2z$. Hence the general solution is

$$x = 0, \qquad y = 2t, \qquad z = 5t,$$

where t is an arbitrary number.

4.7. THEOREM 21. *Given r equal fractions,*

$$\frac{N_1}{D_1} = \frac{N_2}{D_2} = \dots = \frac{N_r}{D_r} \quad (N_1 \neq 0),$$

and any homogeneous function $f(x_1, x_2,\dots, x_r)$ of degree n in the r variables $x_1, x_2,\dots, x_r$, then

$$\left(\frac{N_1}{D_1}\right)^n = \dots = \left(\frac{N_r}{D_r}\right)^n = \frac{f(N_1, N_2,\dots, N_r)}{f(D_1, D_2,\dots, D_r)},$$

provided that $f(D_1, D_2,\dots, D_r)$ is not zero.

Further, if one of the two functions $f(N_1, N_2,\dots, N_r), f(D_1, D_2,\dots, D_r)$ is zero, then so is the other.

Before going on to the proof of this theorem we give, as illustrations, one or two examples.

EXAMPLE 1. When $\dfrac{a}{b} = \dfrac{c}{d}$,

$$\left(\frac{a}{b}\right)^2 = \frac{a^2+c^2}{b^2+d^2} = \frac{ac}{bd} = \frac{la^2+mac+nc^2}{lb^2+mbd+nd^2}.$$

EXAMPLE 2. Let $x_1^2+y_1^2+z_1^2 = 1$, and let

$$\frac{x_1}{l} = \frac{y_1}{m} = \frac{z_1}{n}.$$

Then
$$\frac{x_1}{l} = \frac{y_1}{m} = \frac{z_1}{n} = \pm\sqrt{\frac{x_1^2+y_1^2+z_1^2}{l^2+m^2+n^2}}$$
$$= \pm\frac{1}{\sqrt{(l^2+m^2+n^2)}}.$$

This type of application of Theorem 21 is very widely used in analytical geometry.

EXAMPLE 3. Let $3bd-2f^2 = 0$, and let

$$\frac{a}{b} = \frac{c}{d} = \frac{e}{f}.$$

Then also $3ac-2e^2 = 0$.

4.8. *Proof of Theorem* 21. Let each fraction be equal to t. Then $N_1 = D_1 t$, $N_2 = D_2 t$,..., $N_r = D_r t$, and $t \neq 0$. Thus

$$f(N_1, N_2,..., N_r) = f(D_1 t, D_2 t,..., D_r t)$$

and this (Theorem 17, Corollary) is equal to

$$t^n f(D_1, D_2,..., D_r).$$

Hence
$$f(N_1, N_2,..., N_r) = t^n f(D_1, D_2,..., D_r)$$

and the two parts of the theorem follow.

4.9. *A general method for equal fractions.* The reader will have noticed that the proofs of all our results in § 4 have been obtained by equating each of the equal fractions to a new variable t. It is often useful to take this step without reference to any formal theorems. For example, to solve the problem of § 4.6, (i). Let

$$\frac{x-y+z}{3} = \frac{y-z+x}{4} = \frac{z-x+y}{5} = t.$$

Then

$$x-y+z = 3t, \quad \text{(a)}$$

$$x+y-z = 4t, \quad \text{(b)}$$

$$-x+y+z = 5t, \quad \text{(c)}$$

whence $2x = 7t$ [add (a) and (b)], $2y = 9t$ [add (b) and (c)], $2z = 8t$.

EXAMPLES VIII B*

1. Given that $$\frac{y+1}{y-1} = \frac{(1+x)^2-(1-x)^2}{(1+x)^2+(1-x)^2},$$
prove that $y = -(1+x)^2/(1-x)^2$.

2. Given that $$\frac{y+3}{2y-1} = \frac{\sqrt{(1+2x)}+3\sqrt{(1+x)}}{2\sqrt{(1+2x)}-\sqrt{(1+x)}},$$
prove that $y = \sqrt{(1+2x)}/\sqrt{(1+x)}$.

3. Solve the equation
$$\frac{x+2+\sqrt{(x^2+5)}}{x+2-\sqrt{(x^2+5)}} = 7.$$

In working Examples 4–6 use the method of § 4.9.

4. Find $x:y:z$ when
$$\frac{2x+y-2z}{-1} = \frac{2y+z-2x}{6} = \frac{2z+x-2y}{4}.$$

5. Find $x:y:z$ when
$$\frac{3x+2y+z}{10} = \frac{3y+2z+x}{13} = \frac{3z+2x+y}{13}.$$

6. Find $x:y:z$ when
$$\frac{2x+y-z}{3} = \frac{3x+y-2z}{2} = \frac{x+y+z}{6}.$$

7. Find the solutions of the equations
$$\frac{x+2y}{3} = \frac{y+4z}{4} = \frac{z+x}{5}.$$

8. Solve the equations
$$\frac{2x+y}{7} = \frac{2y+z}{10} = \frac{x+z-y}{3} = 1.$$

9. (*Harder.*) Solve the equations
$$\frac{(x+y)^2-z^2}{-3} = \frac{(y+z)^2-x^2}{5} = \frac{(z+x)^2-y^2}{1} = 3.$$

CHAPTER IX*

RATIONAL FUNCTIONS: GRAPHS

1. Limiting approach

1.1. Before we can draw the graph of a rational function of x we must have clear ideas on how the function behaves

(A) when x is numerically large, positive or negative;

(B) when x is nearly equal to a zero of the denominator.

We begin with a simple example, $y = 1/x^3$; we consider (A) in § 1.2 and (B) in § 1.3. Afterwards we draw the graphs of a few rational functions and show how the considerations (A) and (B) apply to them.

1.2. *To discuss the values of y when $y = 1/x^3$ and the values of x are large, positive or negative.* When x is large and positive (e.g. $x = 10^3$ or 10^7) x^3 is also large and positive; y is then small and positive. As x increases indefinitely, through large positive values, y approaches zero through small positive values.

A table of values brings out this fact, e.g.

$x = 10,$	$10^3,$	$10^7;$
$y = 0{\cdot}001,$	$0{\cdot}000000001,$	10^{-21} [· twenty 0's followed by 1].

The general shape of the graph is given by Fig. 13.

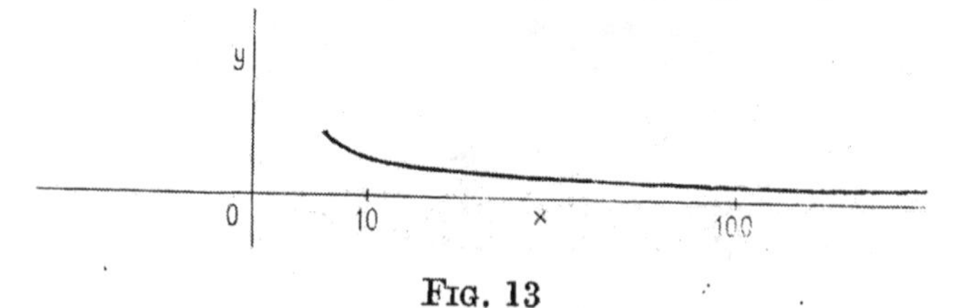

FIG. 13

The curve approaches the line $y = 0$, but lies above it when x is positive.

When x is large and negative, y is small and negative. A table of values shows that y remains negative but approaches zero as the numerical value of x increases; e.g.

$$x = -10, \qquad -10^3, \qquad -10^7;$$
$$y = -1/10^3, \qquad -1/10^9, \qquad -1/10^{21}.$$

The general shape of the graph is given in Fig. 14.

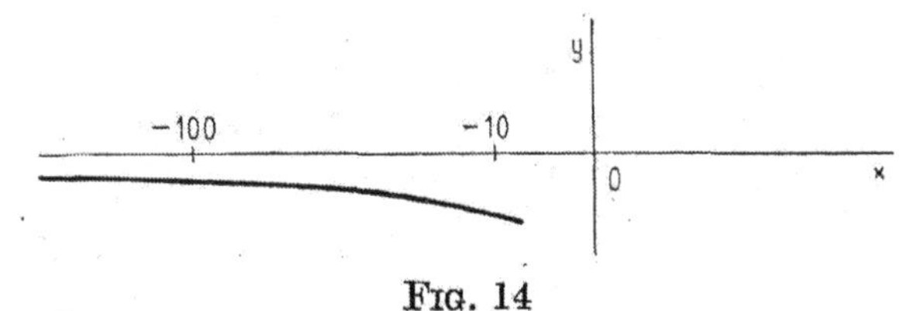

FIG. 14

When x is negative the curve approaches the line $y = 0$, but lies below it.

1.3. *To discuss the values of y when $y = 1/x^3$ and the values of x are numerically small.* When x is small and positive, y is large and positive: e.g.

$$x = 1/10, \qquad 1/10^3, \qquad 1/10^6;$$
$$y = 10^3, \qquad 10^9, \qquad 10^{18}.$$

The values of y increase indefinitely as x approaches the value zero.

When x is small and negative, y is large and negative; e.g.

$$x = -1/10, \qquad -1/10^3, \qquad -1/10^6;$$
$$y = -10^3, \qquad -10^9, \qquad -10^{18}.$$

The general shape of the graph for small values of x is given in Fig. 15.

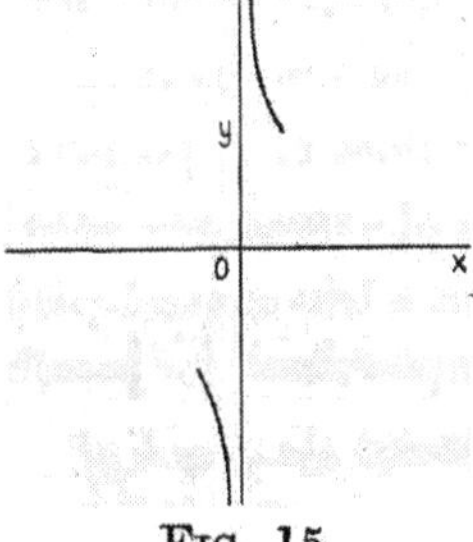

FIG. 15

The curve approaches the line $x = 0$.

1.4. *The rough graph of $y = 1/x^3$.* The general shape of the graph is easily completed. Take one or two control points when

x is neither very small nor very large; e.g.

$$x = 1,\ 4, \qquad -1,\ -4,$$
$$y = 1,\ 0{\cdot}016 \text{ (approx.)},\ -1,\ -0{\cdot}016.$$

The full graph is:

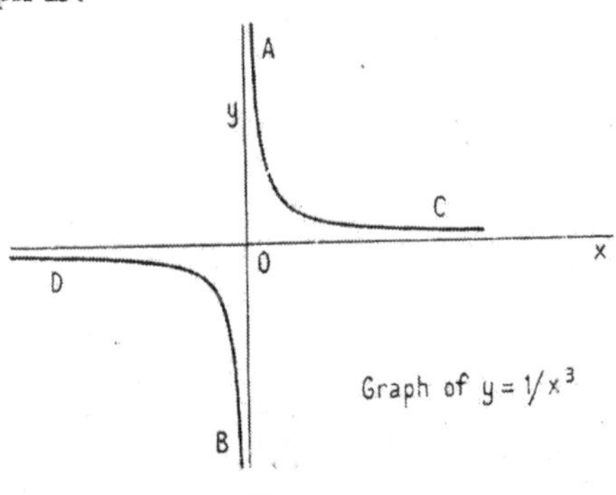

FIG. 16

The curve, as it goes off towards infinity, approaches, without ever quite reaching, the line $x = 0$ (in the parts A and B of the graph) and the line $y = 0$ (in the parts C and D of the graph).

The lines thus approached are called ASYMPTOTES of the curve.

2. Graphs of rational functions

2.1. When we wish to draw the rough graph of a rational function, we use the following procedure:

(i) look for possible turning-points,

(ii) consider large values of x, positive and negative,

(iii) consider values of x near any zero of the denominator,

(iv) *if necessary*, plot a few control-points or even (in some of the more difficult examples) look for possible points of inflexion.

2.2. EXAMPLE 1. *Sketch the graph of*

$$y = \frac{1}{x^2+1}.$$

SOLUTION.

(i)
$$\frac{dy}{dx} = \frac{-2x}{(x^2+1)^2}.$$

This is zero when $x = 0$; dy/dx is positive when x is negative, and is negative when x is positive. There is a maximum at $x = 0$, $y = 1$. Sketch in the part of the curve near this point (point A of the graph).

(ii) When x is large, y is small and y approaches zero as x increases numerically.

When x is large and positive, y is small and positive.

When x is large and negative, y is small and positive.

This shows that the curve lies *above* the line $y = 0$ at both ends and we sketch the parts B and C of the graph.

(iii) The denominator, x^2+1, is never zero when x is real, and so y always remains finite.

(iv) As control-points, take

$$x = -2, \quad -1, \quad 1, \quad 2,$$
$$y = \tfrac{1}{5}, \quad \tfrac{1}{2}, \quad \tfrac{1}{2}, \quad \tfrac{1}{5}.$$

Plot these points and fill in the graph.

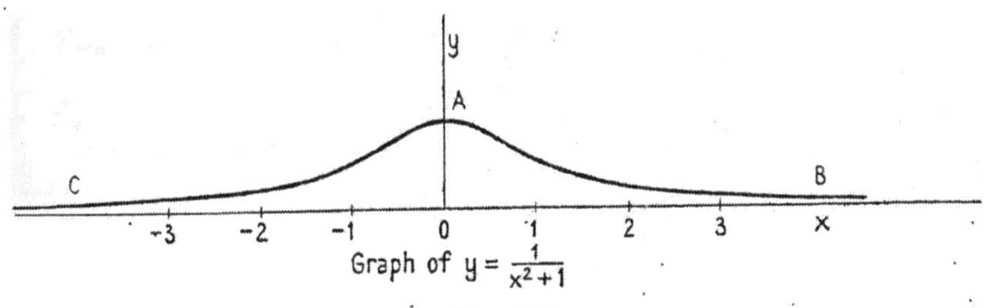

FIG. 17

2.3. EXAMPLE 2. *Sketch the graph of*

$$y = \frac{x-2}{x-3}.$$

SOLUTION.

(i) $$\frac{dy}{dx} = \frac{x-3-(x-2)}{(x-3)^2} = -\frac{1}{(x-3)^2}.$$

Thus, dy/dx is *always negative*; there is no turning-point.

(ii) To consider the values of y when x is large, we divide numerator and denominator by the highest power of x that occurs, here x^1. We obtain in this way

$$y = \frac{x-2}{x-3} = \frac{1-\frac{2}{x}}{1-\frac{3}{x}}.$$

Hence, as x becomes large, y approaches the value 1. Moreover,

$$y-1 = \frac{x-2}{x-3}-1 = \frac{1}{x-3}.$$

Hence, when x is large and positive, y is a little greater than 1; when x is large and negative, y is a little less than 1. The shape of the graph when x is large is

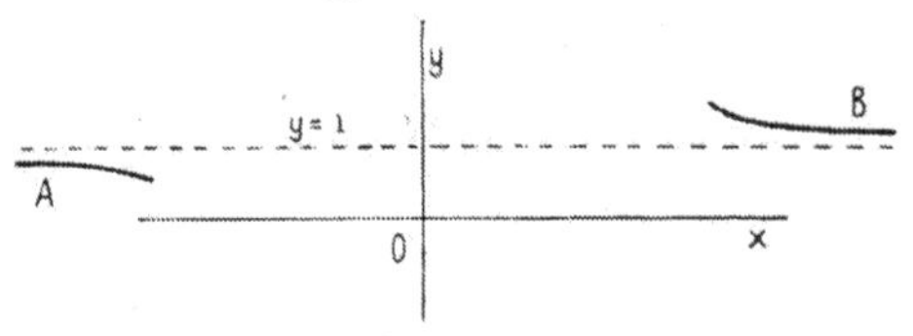

FIG. 18

(iii) The denominator is zero when $x = 3$. Let h be a small positive number.

When $x = 3+h$, $y = \frac{1+h}{h}$, which is large and positive.

When $x = 3-h$, $y = \frac{1-h}{-h}$, which is large and negative.

The shape of the graph when x is nearly equal to 3 is

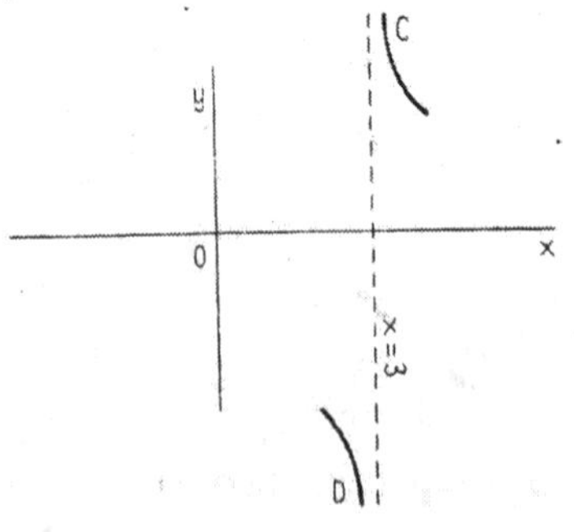

FIG. 19

(iv) As control-points, take

$$x = 0, \quad 4, \quad 2,$$
$$y = \tfrac{2}{3}, \quad 2, \quad 0.$$

These are sufficient to show how the part A joins on to part D and how part B joins on to part C. The full graph is

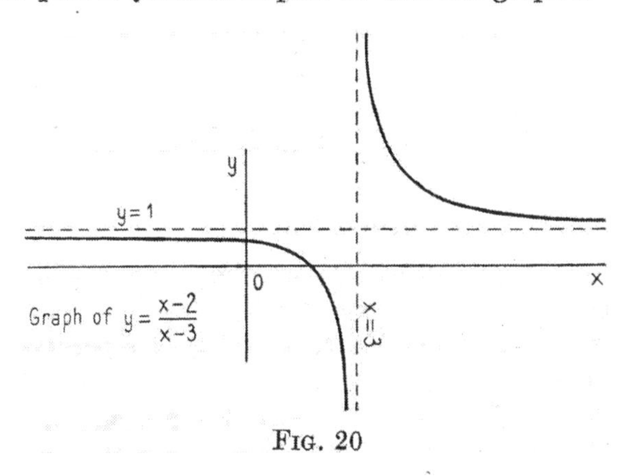

FIG. 20

As the curve goes off to infinity it approaches the lines $x = 3$, $y = 1$, which are therefore asymptotes of the curve.

3. Further examples (*Harder*)

3.1. EXAMPLE 3. *Sketch the graph of*

$$y = \frac{x-1}{(x-2)(x-3)}.$$

SOLUTION.

(i)
$$\frac{dy}{dx} = -\frac{x^2-2x-1}{(x-2)^2(x-3)^2}.$$

$$x^2-2x-1 = 0 \quad \text{when} \quad x = 1 \pm \sqrt{2}.$$

These numbers do not lend themselves to an exact calculation of y: so, unless we want the graph for very precise work, we note that these values are approximately 2·4 and −0·4 and plot the turning-points (2·4, −5·8), (−0·4, −0·2).

(ii) When x is large,†

$$y = \frac{x-1}{x^2-5x+6} = \frac{\frac{1}{x}-\frac{1}{x^2}}{1-\frac{5}{x}+\frac{6}{x^2}} \simeq \frac{1}{x},$$

so that y is positive and small if x is positive, and is negative and small if x is negative. The graph is, for large x,

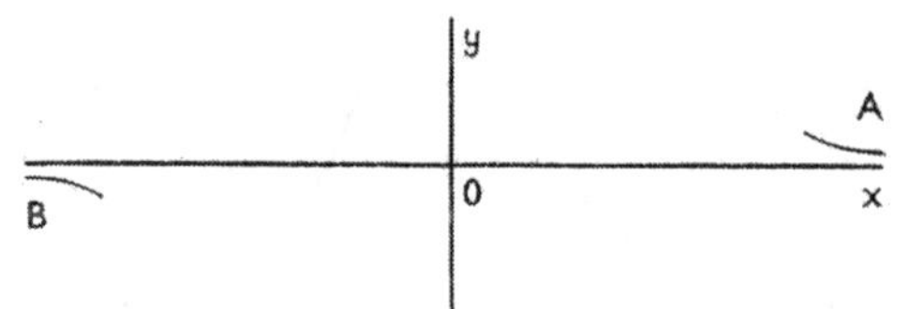

(iii) The zeros of the denominator are $x = 2$, $x = 3$, and y is large when x takes values near 2 or 3.

When $x > 3$, each of $x-1$, $x-2$, and $x-3$ is positive; $y > 0$ since‡ it is $+/+.+$.

When $2 < x < 3$, y is $+/+.-$, and so $y < 0$.

When $1 < x < 2$, y is $+/-.-$, and so $y > 0$.

When $x < 1$, y is $-/-.-$, and so $y < 0$.

These facts show that the approach to the lines $x = 2$, $x = 3$ is given by

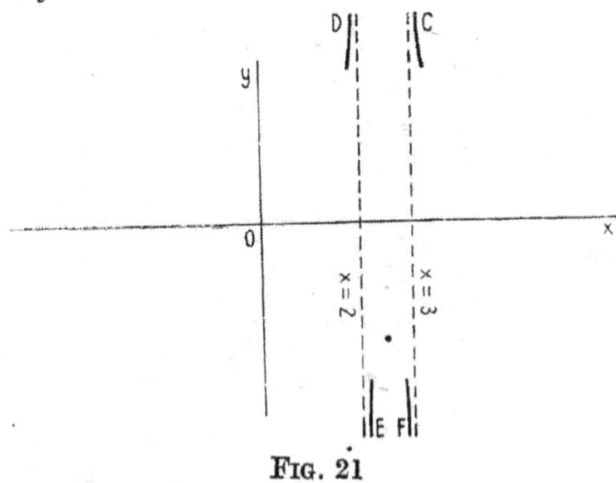

FIG. 21

† The sign $\simeq$ denotes 'is approximately equal to'.

‡ We use $+/+.+$ to denote 'a positive number divided by the product of two positive numbers'; and so for other arrangements of the signs.

(iv) Since $y > 0$ when $x > 3$, C joins with A.

Since $y < 0$ when $2 < x < 3$, E joins with F via the turning-point $x \simeq 2\cdot4$, $y \simeq -5\cdot8$, which we found in (i).

Finally, D must join with B. We need some control-points to see how the join is effected.

There is a turning-point at $x \simeq -0\cdot4$, $y \simeq -0\cdot2$; other obvious points on the curve are

$$x = 1, \quad 0, \quad -1,$$
$$y = 0, \quad -\tfrac{1}{6}, \quad -\tfrac{1}{6}.$$

These points, together with the turning-point, enable us to complete the graph.

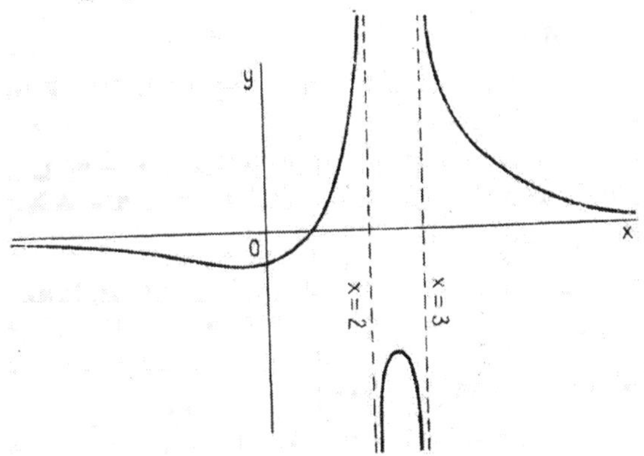

FIG. 22. Graph of $y = \dfrac{x-1}{(x-2)(x-3)}$.

If, for any particular purpose, we needed a more precise drawing of the curve, or any part of it, we should be obliged to plot a larger number of control-points.

3.2. EXAMPLE 4. *Sketch the graph of*

$$y = \frac{(x-1)(2x-1)}{x^2+1}.$$

SOLUTION.†

(i) $$\frac{dy}{dx} = \frac{3x^2+2x-3}{(x^2+1)^2}.$$

† Some simple calculations have been left to the reader. For example, the

The possible turning-points are, approximately,

$$(0{\cdot}7, -0{\cdot}04), \qquad (-1{\cdot}4, 3{\cdot}05).$$

(ii) When x is large,

$$y = \frac{2x^2-3x+1}{x^2+1} = \frac{2-\dfrac{3}{x}+\dfrac{1}{x^2}}{1+\dfrac{1}{x^2}} \simeq 2,$$

and the curve approaches the line $y = 2$.

Moreover,

$$y-2 = -\frac{3x+1}{x^2+1} = \frac{-\dfrac{3}{x}-\dfrac{1}{x^2}}{1+\dfrac{1}{x^2}} \simeq -\frac{3}{x},$$

so that $y > 2$ when x is large and negative, and $y < 2$ when x is large and positive.

This shows that the graph approaches the line $y = 2$ from below out on the right-hand side, and approaches it from above out on the left-hand side.

(iii) The denominator has no real zeros and so y never becomes large.

(iv) Taking the control-points

$$\begin{array}{llllll} x = & -1, & 0, & \tfrac{1}{2}, & 1, & 2, \\ y = & 3, & 1, & 0, & 0, & 0{\cdot}6, \end{array}$$

suitably spaced with reference to the turning-points, we have the graph shown in the figure.

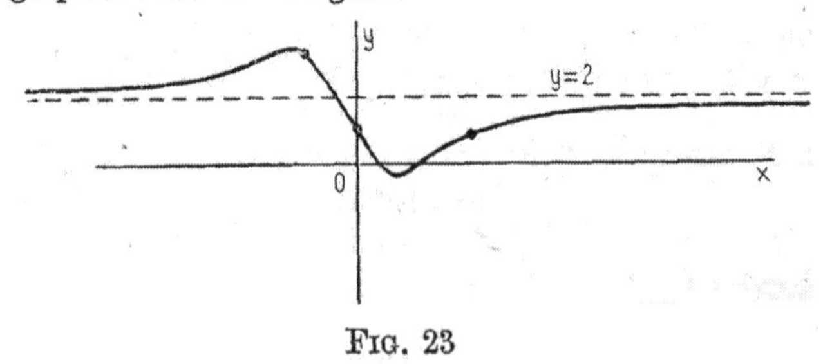

FIG. 23

reader must work out dy/dx for himself: he cannot see how the value is obtained from mere 'inspection'.

4. An application of the theory of quadratic equations

4.1. A glance at the graph of $y = \dfrac{1}{x^2+1}$ in § 2.2 shows that the line $y = k$ will cut the curve in two points if k lies between 0 and 1, but will not cut the curve if k lies outside these limits. The dividing values, $k = 0$ and $k = 1$, fall into neither of the above categories, for $y = 1$ is a tangent to the curve and may be considered to meet it in two coincident points, while $y = 0$ is approached by the curve, which, however, never quite reaches it.

A glance at the graph of $y = (x-2)/(x-3)$ in § 2.3 shows that $y = k$ meets the curve in one point for every value of k except $k = 1$.

We now consider, in one or two particular examples, the question 'Can we find out from the equation of the curve, and without actually drawing the graph, whether y takes all values, as in the graph of $y = (x-2)/(x-3)$, or whether y takes only some values, as in the graphs of §§ 2.2 and 3.2?'

4.2. Example 1. *Show that, for all real values of k other than zero, the line $y = k$ meets the curve*

$$y(x-1)(x-3) = x-2 \tag{1}$$

in two distinct real points.

Solution. Write (1) as an equation in x. We get

$$x^2y-x(4y+1)+3y+2 = 0, \tag{2}$$

which is a quadratic equation in x provided that $y \neq 0$.

The roots of this equation are real and distinct if

$$(4y+1)^2-4y(3y+2) > 0,$$

i.e. if $$4y^2+1 > 0.$$

But this is satisfied for every real value of y.

Hence, whenever y ($\neq 0$) is real, $= k$ say, the equation (2) has two real distinct roots, x_1 and x_2 say, and the line $y = k$ meets the curve in the two points (x_1, k) and (x_2, k).

When $y = 0$, (2) reduces to $x = 2$; and so the line $y = 0$ meets the curve at the one point $(2, 0)$.

EXAMPLE 2. *Prove that no point of the curve*

$$y(x-2)(x-5) = -(x-1) \tag{1}$$

lies between the lines $y = \frac{1}{9}$ *and* $y = 1$.

SOLUTION. Write (1) as an equation in x. We get

$$x^2y-x(7y-1)+10y-1 = 0, \tag{2}$$

which is a quadratic equation unless $y = 0$.

The roots of this equation are not real (Theorem 10) when

$$(7y-1)^2-4y(10y-1) < 0,$$

i.e., when $$9y^2-10y+1 < 0,$$

i.e., when $$(9y-1)(y-1) < 0. \tag{3}$$

This is the case when $\frac{1}{9} < y < 1$, for then the first factor in (3) is positive and the second is negative.

Hence there is no real value of x corresponding to a value of y that lies between $\frac{1}{9}$ and 1. Therefore no part of the graph can lie between the lines $y = \frac{1}{9}$ and $y = 1$.

EXAMPLE 3 (*Harder*). *Prove that x is real for every real value of y if*

$$y = \frac{ax+b}{x^2-1} \quad (a,\ b \text{ real})$$

and $a^2 \geqslant b^2$.

SOLUTION. When $y = 0$, $x = -b/a$ and is real.

When $y \neq 0$, x satisfies the quadratic equation

$$yx^2-ax-(y+b) = 0.$$

Hence x is real if

$$a^2+4y(y+b) \geqslant 0,$$

i.e., if $$4y^2+4by+a^2 \geqslant 0. \tag{1}$$

But (1) may be written as

$$(2y+b)^2+a^2-b^2 \geqslant 0,$$

which is satisfied for all real values of y when $a^2 \geqslant b^2$, and so x is real.

Aliter. We can replace the argument after (1) above by the following. By Theorem 11

$$4y^2+4by+a^2 > 0$$

for all real values of y, if $b^2 < a^2$. Further, when $a^2 = b^2$, the left-hand side of (1) is equal to $(2b+y)^2$, which is positive or zero when y is real. Hence (1) is satisfied if $a^2 \geqslant b^2$.

4.3. NOTE. Most of the examples in which x is to be proved real whenever y is real involve an appeal to Theorem 11. But it is often possible, as in the first proof of Example 3 above, to get the facts by 'completing the square'. All Theorem 11 does is to state in formal terms the results that will follow from considering

$$ax^2+2bx+c \text{ in the form } a\left\{\left(x+\frac{b}{a}\right)^2+\frac{ac-b^2}{a^2}\right\},$$

and when we are working with particular values of a, b, c, it is often easier to complete the square than it is to appeal to Theorem 11.

EXAMPLES IX A

Sketch the graphs of the following:

1. $y-1 = \dfrac{1}{x^3}$, $\quad y+1 = \dfrac{1}{x^3}$.

2. $y = \dfrac{1}{x^2+x+1}$, $\quad y = \dfrac{1}{2x^2+3}$.

3. $y = \dfrac{x}{x^2+1}$, $\quad y = \dfrac{x^2+x+1}{x^2+1}$.

4. $y = \dfrac{x-1}{x-2}$, $\quad y = \dfrac{2x-1}{2x-3}$.

5. $y = \dfrac{x-2}{(x-1)(x-3)}$, $\quad y = \dfrac{x}{x^2-1}$.

6. $y = \dfrac{x-3}{(x-1)(x-2)}$, $\quad y = \dfrac{x}{(x-1)(x-2)}$.

7. $y = \dfrac{x^2-1}{x^2+1}$, $\quad y = \dfrac{x^2-3x+2}{2x^2+1}$.

EXAMPLES IX B

1. Prove that two distinct real values of x correspond to every real value of y other than zero

(i) when $y = \dfrac{x-3}{(x-2)(x-4)}$, $\quad$ (ii) when $y = \dfrac{x-2}{(x-1)(x-4)}$.

2. (i) Prove that no part of the graph of

$$y = x/(x-1)(x-4)$$

lies between the lines $y = -\frac{1}{9}$ and $y = -1$.

(ii) Prove that no part of the graph of

$$y = (x-1)/(x^2+3x)$$

lies between the lines $y = \frac{1}{9}$ and $y = 1$.

3. (i) Prove that no part of the graph of

$$y = (x-3)/(x-2)^2$$

lies above the line $y = \frac{1}{4}$.

(ii) Prove that no part of the graph of

$$y = (x-4)/(x-1)^2$$

lies above the line $y = \frac{1}{12}$.

4. Prove that x is real for every real value of y if

$$y = \frac{x+1}{a^2x^2-b^2}$$

and $b^2 \geqslant a^2$.

5. Prove that no part of the graph of

$$y = \frac{20(x+1)}{25x^2-16}$$

lies between the lines $y = -1$ and $y = -\frac{1}{4}$.

CHAPTER X

THE BINOMIAL THEOREM FOR ANY INDEX

1. The introduction of infinite series

1.1. If we were set the elementary problem of dividing $1-x^5$ by $1-x$ and expressing the quotient in a series of ascending powers of x we should proceed, using the ordinary long division process, as follows:

$$\begin{array}{l}
1-x)1 \qquad\qquad -x^5(1+x+x^2+x^3+x^4 \\
\quad\;\; \underline{1-x} \\
\qquad\quad x \\
\qquad\quad \underline{x-x^2} \\
\qquad\qquad\; x^2 \\
\qquad\qquad\; \underline{x^2-x^3} \\
\qquad\qquad\qquad x^3 \\
\qquad\qquad\qquad \underline{x^3-x^4} \\
\qquad\qquad\qquad\quad x^4-x^5 \\
\qquad\qquad\qquad\quad \underline{\underline{x^4-x^5}}
\end{array}$$

The long division terminates when we arrive at the stage at which there is no remainder: this happens when we bring down the term in x^5.

If now we start to divide 1 by $1-x$ and express the quotient in a series of ascending powers of x, there is no term to bring down and the long-division process will go on for ever. The quotient would appear to be

$$1+x+x^2+\ldots \textit{ ad infinitum}. \qquad (1)$$

We may therefore ask ourselves whether there is any way of interpreting the series (1) so that the result makes sense. The reader will see for himself that simply to say 'Take 1, add x, add x^2, add x^3, and go on like this for ever' is just nonsense. If we are to give (1) a meaning at all, that meaning must be expressed by means of a property belonging to the sum of a finite number of the terms of (1): we can find the sum of a finite number of things and we cannot give a precise meaning to 'adding up an infinite number of things'.

Now the sum of n terms of (1) is

$$1+x+x^2+...+x^{n-1}, \tag{2}$$

which is equal to

$$\frac{1-x^n}{1-x},$$

or, as we write it with a view to our next step,

$$\frac{1}{1-x}-\frac{x^n}{1-x}. \tag{3}$$

1.2. *A convergent series.* Suppose first that $x \neq 0$, but has a value lying between -1 and $+1$. Then the number x^n approaches zero as n increases; the following table will illustrate this.†

x	x^{10}	x^{20}	x^{30}	x^{100}
0·1	10^{-10}	10^{-20}	10^{-30}	10^{-100}
0·5	10^{-3}	10^{-6}	10^{-9}	8.10^{-31}
0·9	0·35	0·12	0·04	0·00003

Thus, in the expression (3) of § 1.1,

$$\frac{x^n}{1-x}$$

approaches zero as n increases, and so

$$\frac{1}{1-x}-\frac{x^n}{1-x} \quad \text{approaches the value} \quad \frac{1}{1-x}.$$

This shows that the sum of the first n terms of the infinite series

$$1+x+x^2+... \textit{ ad infinitum} \tag{4}$$

approaches the value $1/(1-x)$ as n increases. In this sense we may use the infinite series (4) to represent $1/(1-x)$, and we write

$$\frac{1}{1-x} = 1+x+x^2+... \textit{ ad inf.}$$

We say that the series 'converges to $1/(1-x)$' and we call the series 'a convergent series'.

1.3. *A series that does not converge.* Suppose next that $x > 1$ or that $x < -1$; for example $x = 2$ or $x = -10$. Then the numerical value of x^n increases beyond all bounds as n increases; the table on p. 127 will illustrate this.

† Some of the values given are rough approximations only.

x	x^{10}	x^{20}	x^{30}	x^{100}
2	10^3	10^6	10^9	10^{30} (approx.)
-10	10^{10}	10^{20}	10^{30}	10^{100}

The more we increase n, the more the value of $1+x+...+x^{n-1}$, that is, of

$$\frac{1}{1-x}-\frac{x^n}{1-x},$$

differs from $1/(1-x)$. The sum of the first n terms of the infinite series

$$1+x+x^2+...\ ad\ inf.$$

does not approach the value of $1/(1-x)$. We cannot use the series to represent the function $1/(1-x)$.

1.4. *Summary.* When $-1 < x < 1$ and $x \neq 0$, the sum of n terms of the series

$$1+x+x^2+...\ ad\ inf.$$

approaches the value $1/(1-x)$, and in this sense we may write

$$\frac{1}{1-x} = 1+x+x^2+...\ ad\ inf. \tag{5}$$

When $x = 0$, the series $1+0+0+...$ to n terms has the sum 1 for all values of n; the value of $1/(1-x)$ is also 1: the two are equal. We can legitimately use (5) when $x = 0$; but there is, of course, not much point in so doing, and we mention the matter here chiefly to show what happens when the value $x = 0$, which we excluded previously, is taken into account.

When $x < -1$ or when $x > 1$ we cannot legitimately use equation (5); for then the sum of n terms of $1+x+x^2+...$ does not approach $1/(1-x)$.

2. Further examples of infinite series

2.1. *The square root of* $1+x$. The elementary process of finding the square root of $1+x$ gives, as its first few steps,

$$\begin{array}{r|l}
1 & 1+x\,(1+\tfrac{1}{2}x-\tfrac{1}{8}x^2+\tfrac{1}{16}x^3+... \\
 & 1 \\
\hline
2+\tfrac{1}{2}x & x \\
 & x+\tfrac{1}{4}x^2 \\
\hline
2+x-\tfrac{1}{8}x^2 & -\tfrac{1}{4}x^2 \\
 & -\tfrac{1}{4}x^2-\tfrac{1}{8}x^3+\tfrac{1}{64}x^4 \\
\hline
2+x-\tfrac{1}{4}x^2+\tfrac{1}{16}x^3 & \quad \tfrac{1}{8}x^3-\tfrac{1}{64}x^4
\end{array}$$

The process goes on for ever, and it suggests, in the light of § 1, that, for some values of x at any rate, $(1+x)^{\frac{1}{2}}$ can be represented by the infinite series

$$1+\tfrac{1}{2}x-\tfrac{1}{8}x^2+\tfrac{1}{16}x^3-\ldots. \tag{1}$$

The law governing the formation of the coefficients is not easy to see from this way of getting at the series. But if we look at (1) in relation to the binomial theorem (as we proved it in Chapter VI), we see the law without any difficulty.

When n is a *positive integer* we know that (Chapter VI, Theorem 16, Note 4)

$$(1+x)^n = 1+nx+\frac{n(n-1)}{2!}x^2+\frac{n(n-1)(n-2)}{3!}x^3+\ldots.$$

Let us put $n = \dfrac{1}{2}$ in the series on the right. Then

$$n = \frac{1}{2}; \qquad \frac{n(n-1)}{2!} = \frac{\frac{1}{2}(-\frac{1}{2})}{2} = -\frac{1}{8};$$

$$\frac{n(n-1)(n-2)}{3!} = \frac{\frac{1}{2}(-\frac{1}{2})(-\frac{3}{2})}{3.2} = \frac{1}{16}.$$

These are the numbers given as the coefficients of x, x^2, and x^3 by the square-root process. It is indicated that

$$(1+x)^{\frac{1}{2}} \text{ is represented by } 1+\tfrac{1}{2}x+\frac{\frac{1}{2}(\frac{1}{2}-1)}{2!}x^2+\ldots,$$

the series continuing *ad infinitum*.

2.2. *The series for* $(1-x)^{-1}$. Again, let us test what happens when we put $n = -1$ and $-x$ for x in the series

$$1+nx+\frac{n(n-1)}{2!}x^2+\frac{n(n-1)(n-2)}{3!}x^3+\ldots. \tag{2}$$

We get

$$1-(-x)+\frac{(-1)(-2)}{1.2}x^2-\frac{(-1)(-2)(-3)}{3.2.1}x^3+\ldots$$

$$= 1+x+x^2+x^3+\ldots. \tag{3}$$

That is to say, the series (3) is given by using the binomial expansion (2) with $n = -1$ and with $-x$ for x; and it represents $(1-x)^{-1}$ when $-1 < x < 1$.

2.3. Thus we have discovered two examples, one with $n = \frac{1}{2}$ (a fraction) and one with $n = -1$ (a negative number), in which the infinite series

$$1+nx+\frac{n(n-1)}{2!}x^2+\frac{n(n-1)(n-2)}{3!}x^3+\ldots$$

represents the function $(1+x)^n$.

Other examples are readily obtained,† though more than a modicum of patience is required to make the calculations beyond the first few terms. There is no point in multiplying the number of examples.

All the examples are particular cases of a general theorem, which is of the first importance in mathematical work. We shall give a formal statement of this theorem.

THEOREM 22. THE BINOMIAL THEOREM. *Let n be any number, positive or negative, integral or fractional. Then the function* $(1+x)^n$ *is represented by the binomial series*

$$1+nx+\frac{n(n-1)}{2!}x^2+\ldots+\frac{n(n-1)\ldots(n-r+1)}{r!}x^r+\ldots$$

whenever $-1 < x < 1$.

NOTES ON THE THEOREM.

NOTE 1. When n *is a positive integer* the series terminates (as in Theorem 16) and its sum is $(1+x)^n$ *for all values of* x.

NOTE 2. When n is not a positive integer (or zero) the series is an infinite series and it represents the function $(1+x)^n$ when $-1 < x < 1$; for other values of x it does not, save in some very special cases, represent the function.

NOTE 3. When n is a fraction, say $n = p/q$, where p is an integer (positive or negative) and q is a positive integer, the

† The reader who has the curiosity to see for himself should try

(*a*) by long division, $\frac{1}{(1-x)^2} = \frac{1}{1-2x+x^2} = 1+2x+3x^2+4x^3+\ldots$,

(*b*) by square root,

$$(1+x)^{\frac{3}{2}} = \sqrt{(1+3x+3x^2+x^3)} = 1+\tfrac{3}{2}x+\tfrac{3}{8}x^2-\tfrac{1}{16}x^3+\ldots.$$

function represented by the binomial series is 'the positive qth root of $(1+x)^p$'. For example,

$$1+\tfrac{1}{2}x+\frac{\frac{1}{2}(\frac{1}{2}-1)}{2!}x^2+\frac{\frac{1}{2}(\frac{1}{2}-1)(\frac{1}{2}-2)}{3!}x^3+\dots$$

represents $+\sqrt{(1+x)}$ and

$$1+\tfrac{5}{4}x+\frac{\frac{5}{4}(\frac{5}{4}-1)}{2!}x^2+\dots$$

represents $+\sqrt[4]{(1+x)^5}$.

NOTE 4. We stress again the sense in which we must interpret the equation

$$(1+x)^n = 1+nx+\frac{n(n-1)}{2!}x^2+\dots\ ad\ inf.\quad (-1<x<1)$$

when n is not a positive integer.

The equation means, not that we can add up an infinite number of terms, but that the sum of M terms of the series on the right approaches the value $(1+x)^n$ as M increases.

The series is often called the EXPANSION of $(1+x)^n$.

2.4. *The proof of Theorem* 22. A proof of Theorem 22 is beyond the range of this book; it requires a knowledge of the general theory of infinite series, and in this book we do not intend to go into the details of this subject.

The reader can use the infinite series with complete confidence provided always that he remembers the essential point stressed in Note 4 above.

3. Applications of the binomial theorem

3.1. *The general term.* By the theorem, the coefficient of x^r in the expansion of $(1+x)^n$ is

$$\frac{n(n-1)(n-2)\dots(n-r+1)}{r!}.$$

This form of the coefficient is often capable of considerable reduction. We work out a few examples.

EXAMPLE 1. *To find the coefficient of x^r in the expansion of* $(1+x)^{-1}$.

The coefficient of x^r is

$$\frac{(-1)(-2)\dots(-1-r+1)}{r!} = \frac{(-1)(-2)\dots(-r)}{r!}.$$

There are r factors in the numerator, and so the coefficient is

$$(-1)^r.\frac{1.2\dots r}{r!} = (-1)^r.$$

Thus, when $-1 < x < 1$,

$$(1+x)^{-1} = 1-x+x^2-\dots+(-1)^r x^r+\dots,$$

and $$(1-x)^{-1} = 1+x+x^2+\dots+x^r+\dots.$$

EXAMPLE 2. *To find the coefficient of x^r in the expansion of* $(1+x)^{-2}$.

The coefficient of x^r is

$$\frac{(-2)(-3)\dots(-2-r+1)}{r!} = (-1)^r.\frac{2.3\dots(r+1)}{1.2\dots r}$$
$$= (-1)^r(r+1).$$

Thus, when $-1 < x < 1$,

$$(1+x)^{-2} = 1-2x+3x^2-\dots+(-1)^r(r+1)x^r+\dots,$$

and $$(1-x)^{-2} = 1+2x+3x^2+\dots+(r+1)x^r+\dots.$$

EXAMPLE 3. *To find the coefficient of x^r in the expansion of* $(1-x)^{-\frac{1}{2}}$.

The term in x^r is

$$\frac{(-\frac{1}{2})(-\frac{1}{2}-1)(-\frac{1}{2}-2)\dots(-\frac{1}{2}-r+1)}{r!}(-x)^r$$
$$= (-1)^r.\frac{1.3.5\dots(2r-1)}{2^r.1.2.3\dots r}(-1)^r x^r$$
$$= \frac{1.3.5\dots(2r-1)}{2.4.6\dots 2r}x^r.$$

Thus, when $-1 < x < 1$,

$$(1-x)^{-\frac{1}{2}} = 1+\frac{1}{2}x+\frac{1.3}{2.4}x^2+\dots+\frac{1.3\dots(2r-1)}{2.4\dots 2r}x^r+\dots.$$

3.2. *Functions like* $(2+x)^n$, $(1+3x)^n$.

(i) In Theorem 22 it is $(1+x)^n$ that is in question and in order to be able to apply the theorem to the expansion of

$(2+x)^n$ we must first write that function in the form $2^n(1+\frac{1}{2}x)^n$. We then get

$$(2+x)^n = 2^n(1+\tfrac{1}{2}x)^n$$
$$= 2^n\left\{1+n.\frac{x}{2}+\frac{n(n-1)}{2!}\frac{x^2}{2^2}+\dots\ ad\ inf.\right\},$$

the coefficient of x^r in this being

$$\frac{n(n-1)\dots(n-r+1)}{r!}2^{n-r}.$$

Moreover, we can use Theorem 22 to represent $(1+\frac{1}{2}x)^n$ by the infinite series whenever $-1 < \frac{1}{2}x < 1$; that is, whenever $-2 < x < 2$. As a particular case, we note

$$\frac{1}{(2+x)} = \frac{1}{2(1+\frac{1}{2}x)} = \frac{1}{2}\left(1-\frac{1}{2}x+\frac{1}{2^2}x^2-\dots\right).$$

(ii) When we use Theorem 22 to represent $(1+3x)^n$ by an infinite series, we get

$$(1+3x)^n = 1+n.3x+\frac{n(n-1)}{2!}3^2x^2+\dots\ ad\ inf.,$$

the equality sign being justified when $-1 < 3x < 1$; that is, when $-\frac{1}{3} < x < \frac{1}{3}$.

3.3. *Rational functions represented by infinite series.*

EXAMPLE 1. *Find the expansion of*

$$\frac{1+x}{(1-2x)(1-x)^2}$$

in a series of ascending powers of x, giving the first few terms of the series and the general term. For what values of x does the result hold?

SOLUTION. By the method of partial fractions we find (the details are left to the reader)

$$\frac{1+x}{(1-2x)(1-x)^2} \equiv \frac{6}{1-2x}-\frac{3}{1-x}-\frac{2}{(1-x)^2}. \quad (1)$$

By the Binomial Theorem,

$$6(1-2x)^{-1} = 6\{1+2x+2^2x^2+...+2^rx^r+...\},$$

$$3(1-x)^{-1} = 3\{1+x+x^2+...+x^r+...\},$$

$$2(1-x)^{-2} = 2\{1+2x+3x^2+...+(r+1)x^r+...\}.$$

Thus (1) is equal to

$$1+5x+15x^2+...+(6.2^r-3-2r-2)x^r+...$$
$$= 1+5x+15x^2+...+(6.2^r-2r-5)x^r+.... \quad (2)$$

The series for $(1-2x)^{-1}$ holds good when $-1 < 2x < 1$, i.e. when $-\frac{1}{2} < x < \frac{1}{2}$; the series for $(1-x)^{-1}$ and $(1-x)^{-2}$ hold good when $-1 < x < 1$. Hence the series (2) holds when $-\frac{1}{2} < x < \frac{1}{2}$.

3.4. *Approximations.* When $-1 < x < 1$, the sum of r terms of the series

$$1+nx+\frac{n(n-1)}{2!}x^2+... \quad (1)$$

approaches, as r increases, the value of $(1+x)^n$. The series (1) therefore provides us with a means of approximating to that value. In general, the larger number of terms we take, the closer is the approximation.

There are two main types of approximation for which the series can be used:

(A) numerical approximations, such as finding the value of $(1{\cdot}1)^{-3}$ correct to 4 significant figures;

(B) approximations correct to a given power of x, such as finding an approximation for $(1+4x)^{-\frac{1}{2}}$ correct to terms in x^2.

Of these (A) is straightforward numerical calculation of the type considered in Chapter VI, § 5.2; and (B) is based on the ideas explained in Chapter VI, § 5.5. We shall work one example of each kind.

EXAMPLE 1. *Find* $(1{\cdot}1)^{-3}$ *correct to 3 significant figures.*

SOLUTION. Since $0{\cdot}1$ lies between -1 and $+1$,

$$(1{\cdot}1)^{-3} = (1+\tfrac{1}{10})^{-3}$$

$$= 1-3.\frac{1}{10}+\frac{(-3)(-4)}{2}.\frac{1}{10^2}+\frac{(-3)(-4)(-5)}{3.2}.\frac{1}{10^3}+$$

$$+\frac{(-3)(-4)(-5)(-6)}{4.3.2}.\frac{1}{10^4}+\dots$$

$$= 1-\frac{3}{10}+\frac{6}{10^2}-\frac{1}{10^2}+\frac{15}{10^4}-\frac{21}{10^5}+\dots$$

$$\begin{array}{lcl} +1{\cdot}0 & \qquad & -0{\cdot}3 \\ 0{\cdot}06 & & 0{\cdot}01 \\ 0{\cdot}0015 & & 0{\cdot}0002 \\ \hline 1{\cdot}0615 & & 0{\cdot}3102 \\ 0{\cdot}3102 & & \\ \hline 0{\cdot}7513 & & \end{array}$$

The value to 3 significant figures is $0{\cdot}751$.

EXAMPLE 2. *Find the expansion of*

$$\frac{1}{1+x+x^2}$$

correct to terms in x^3.

SOLUTION (*a*). By the Binomial Theorem,

$$(1+y)^{-1} = 1-y+y^2-y^3+\dots,$$

and so, on writing $y = x+x^2 = x(1+x)$,

$$(1+x+x^2)^{-1} = 1-x(1+x)+x^2(1+x)^2-x^3(1+x)^3+\dots.$$

Neglecting terms in x^4, x^5,..., we have†

$$(1+x+x^2)^{-1} \simeq 1-x-x^2+x^2(1+2x)-x^3.1$$

$$= 1-x+x^3.$$

† Notice that we write $x^2(1+x)^2 \equiv x^2(1+2x+x^2)$

as $$x^2(1+2x).$$

We omit the term that would lead to an x^4.

Again, we write $x^3(1+x)^3 \equiv x^3(1+3x+3x^2+x^3)$

as $$x^3.1.$$

We omit the terms that would lead to x^4, x^5, or x^6.

SOLUTION (*b*).

$$\frac{1}{1+x+x^2} = \frac{1-x}{1-x^3} = (1-x)(1-x^3)^{-1}.$$

On using the expansion

$$(1-y)^{-1} = 1+y+y^2+\dots$$

with $y = x^3$, and omitting all powers of x higher than the third,

$$\frac{1}{1+x+x^2} \simeq (1-x)(1+x^3) \simeq 1-x+x^3.$$

4. Particular expansions

Some applications of the binomial theorem are so frequent that it is worth while to remember the expansion in the form it takes after the coefficients have been expressed in the simplest manner possible. Notable examples are

$$(1-x)^{-1} = 1+x+x^2+\dots$$

$$(1-x)^{-2} = 1+2x+3x^2+\dots+(r+1)x^r+\dots.$$

It is certainly worth while to be able to write these down without having to work them out from Theorem 22.

Other series that one should be able to recognize when they occur are

$$(1-x)^{-\frac{1}{2}} = 1+\frac{1}{2}x+\frac{1.3}{2.4}x^2+\dots+\frac{1.3\dots(2r-1)}{2.4\dots2r}x^r+\dots$$

$$(1-x)^{\frac{1}{2}} = 1-\frac{1}{2}x-\frac{1}{2.4}x^2-\dots-\frac{1.3\dots(2r-3)}{2.4\dots2r}x^r-\dots$$

$$(1-x)^{-3} = 1+3x+\frac{3.4}{2}x^2+\dots+\frac{(r+1)(r+2)}{2}x^r+\dots.$$

It is not worth while to learn these thoroughly unless one has the flair for memorizing such things easily and, above all, accurately.

EXAMPLES X

1. Find the expansions of

(i) $(1-x)^{-3}$, (ii) $(1+x)^{\frac{1}{2}}$,

giving the first 4 terms and the general term of each expansion.

2. Find the coefficients of x^3 and of x^5 in the expansions of

$$\text{(i)}\quad (1+2x)^{-\frac{3}{2}}, \qquad \text{(ii)}\quad (1+3x)^{-\frac{1}{3}},$$

and find, in their simplest forms, expressions for the coefficients of x^k in these expansions. For what values of x is each expansion valid?

3. Find the coefficient of x^r in the expansion of

$$\text{(i)}\quad \frac{x}{(1-x)^2}, \qquad \text{(ii)}\quad \frac{x^2}{1-x}.$$

4. The functions

$$\text{(i)}\quad \frac{1+x}{1-x}, \qquad \text{(ii)}\quad \frac{1}{(1-x)^3}-\frac{1}{1-x}$$

are expanded in a series of ascending powers of x. Find the first 3 terms and the general term of each expansion.

5. Find the coefficient of x^r in the expansion of

$$\text{(i)}\quad (2+x)^n, \qquad \text{(ii)}\quad (2+3x)^n, \qquad \text{(iii)}\quad (3-2x)^n.$$

For what values of x is each expansion valid?

6. Find the first 3 terms and the general term in the expansion of

$$\text{(i)}\quad \frac{1}{(2+x)^2}, \qquad \text{(ii)}\quad \frac{1}{\sqrt{(2-x)}}, \qquad \text{(iii)}\quad (3-x)^{-3}.$$

and state the range of values of x for which the expansion holds good.

In each of the examples 7–9 find the expansion of the function in a series of ascending powers of x, giving the first 3 terms and the general term. State the values of x for which the expansion is valid.

In each example first express the given rational function as a sum of partial fractions and then use the binomial theorem (*vide* § 3.3).

7. $$\text{(i)}\quad \frac{1}{(1-x)(1-2x)}, \qquad \text{(ii)}\quad \frac{1}{(1+x)(1+2x)}.$$

8. $$\text{(i)}\quad \frac{2-3x}{(1-x)(1-2x)}, \qquad \text{(ii)}\quad \frac{2-x}{(1-x)^2}.$$

9. $$\text{(i)}\quad \frac{1+x^2}{(1-x)^3}, \qquad \text{(ii)}\quad \frac{x^2}{(1-x)^2(1-2x)}.$$

10. Find to 3 significant figures the values of

$$\text{(i)}\quad (1{\cdot}01)^{-9}, \qquad \text{(ii)}\quad (1{\cdot}05)^{-\frac{3}{2}}.$$

Use the method of § 3.4, Example 1, and check your answer against the result as given by the use of four-figure tables of logarithms.

11. Find to 4 significant figures the values of

$$\text{(i)}\quad (9{\cdot}99)^{-6}, \qquad \text{(ii)}\quad (99{\cdot}99)^{\frac{3}{2}}.$$

NOTE. $$(9{\cdot}99)^{-6} = \left(10-\frac{1}{10^2}\right)^{-6} = 10^{-6}\left(1-\frac{1}{10^3}\right)^{-6}.$$

Check your result against the answer given by the use of four-figure tables of logarithms.

12. Find the expansion correct to terms in x^3 of

$$\text{(i)}\quad \frac{1+x}{1-3x}, \qquad \text{(ii)}\quad \frac{1}{1-2x+3x^2}.$$

13. Find the expansion correct to terms in x^3 of

$$\text{(i)}\quad (1+x+x^2+x^3)^{-5}, \qquad \text{(ii)}\quad (1+2x)^{-3}(1+3x)^2.$$

14. (*Harder.*) Find the expansion correct to terms in x^3 of

$$\text{(i)}\quad (2+3x)^{-7}, \qquad \text{(ii)}\quad (4+2x+x^2)^{-\frac{3}{2}}.$$

15. Prove that, when x is small,

$$\frac{1}{1-x} \simeq 1+x, \qquad \sqrt{(1+x)} \simeq 1+\tfrac{1}{2}x,$$

$$(1+3x)^{-\frac{1}{3}} \simeq 1-x, \qquad (1+2x)^{\frac{3}{2}} \simeq 1+3x.$$

16. In the expansion of $a(1-x)^{-2}+b(1+x)^{-1}$, the coefficient of x^6 is 18 and the coefficient of x^7 is 12. Find the values of a and b.

17. $a(1+x)+b(1+x)^2+c(1+x)^3 \simeq 5+(1-2x)^{-1}$, the approximation being correct to terms in x^3. Find the values of a, b, c.

18. (*Harder.*) The expansion in ascending powers of x of

$$\frac{a}{1-x}+\frac{b}{(1-x)^2}+\frac{c}{(1-x)^3}$$

begins with the term $1.x^2$, there being no constant term and the coefficient of x being zero. Prove that the coefficient of x^r in the expansion is $\frac{1}{2}r(r-1)$.

CHAPTER XI

THE EXPONENTIAL AND LOGARITHMIC FUNCTIONS

Introduction

The use of infinite series introduces functions that do not belong to the study of algebra. But the more important of these functions are closely related to algebra and it is usual to give some account of them in an algebra book. These functions are the exponential function and the logarithmic function; they are used in almost every branch of mathematics.

To discuss the functions adequately a fairly extensive knowledge of the theory of infinite series is necessary. Here we shall attempt no more than a sketch of one of the many possible ways of introducing the functions.

1. The exponential function

1.1. In general, when y is a function of x, the differential coefficient dy/dx is not identically equal to y. For example,

$$\text{when } y = x^3, \qquad \frac{dy}{dx} = 3x^2;$$

and $3x^2$ is not identically equal to x^3.

The particular function y whose differential coefficient is identically equal to y is found to be represented by the infinite series

$$y = 1+x+\frac{x^2}{2!}+\frac{x^3}{3!}+\ldots+\frac{x^n}{n!}+\ldots. \tag{1}$$

Assuming (what can in fact be proved) that the differential coefficient is given correctly by differentiating each term of the series separately, we have

$$\frac{dy}{dx} = 0+1+x+\frac{x^2}{2!}+\ldots+\frac{x^{n-1}}{(n-1)!}+\ldots. \tag{2}$$

Since the series in (1) and (2) are the same,

$$y \equiv \frac{dy}{dx}.$$

The value of (1) when $x = 1$ is

$$1+1+\frac{1}{2!}+\frac{1}{3!}+\dots. \tag{3}$$

The sum of the first n terms of this series approaches a certain definite value as n increases. This definite value is denoted by e. The numerical value is approximately† 2·7183, which is obtained by taking the first 9 terms of the series (3).

The series (1) above can be proved (by the theory of infinite series) to be equal to e^x, so that

$$e = 1+1+\frac{1}{2!}+\frac{1}{3!}+\dots, \tag{4}$$

and
$$e^x = 1+x+\frac{x^2}{2!}+\frac{x^3}{3!}+\dots. \tag{5}$$

Before proceeding further we stress the meaning to be attached to the equation (5). It means that, for any given value of x,

$$1+x+\frac{x^2}{2!}+\dots+\frac{x^n}{n!}$$

approaches the value $(e)^x$ as n increases. Unlike the binomial theorem, THIS IS TRUE FOR ALL VALUES OF x.

1.2. The reader should be prepared to recognize this series when x is replaced by simple functions of x, such as $-x^2$ or $3x$. For example,

$$e^{-x} = 1-x+\frac{x^2}{2!}-\frac{x^3}{3!}+\dots+(-1)^n\frac{x^n}{n!}+\dots,$$

$$e^{2x} = 1+2x+\frac{4x^2}{2!}+\frac{8x^3}{3!}+\dots+\frac{(2x)^n}{n!}+\dots,$$

$$e^{-x^2} = 1-x^2+\frac{x^4}{2!}-\frac{x^6}{3!}+\dots+(-1)^n\frac{x^{2n}}{n!}+\dots.$$

Again, the theory of infinite series proves that we can add or subtract such series when they are convergent, and examples of this are of common occurrence. For example,

$$e^x = 1+x+\frac{x^2}{2!}+\frac{x^3}{3!}+\dots, \qquad e^{-x} = 1-x+\frac{x^2}{2!}-\frac{x^3}{3!}+\dots.$$

† The value has been calculated to an enormous number of places of decimals; the first 10 places give 2·7182818285. It is not a recurring decimal.

By addition and division by 2,

$$\tfrac{1}{2}(e^x+e^{-x}) = 1+\frac{x^2}{2!}+\frac{x^4}{4!}+\dots+\frac{x^{2n}}{(2n)!}+\dots,$$

while, by subtraction and division by 2,

$$\tfrac{1}{2}(e^x-e^{-x}) = x+\frac{x^3}{3!}+\dots+\frac{x^{2n+1}}{(2n+1)!}+\dots.$$

2. The logarithmic function

2.1. In numerical calculations logarithms are usually taken to the base 10. In theoretical work logarithms are taken to the base e.

We assume that the reader is familiar with the facts briefly outlined in what follows. Let logarithms be taken to any base a. Let

$$x = a^n, \qquad y = a^m. \tag{1}$$

Then $\log x = n, \qquad \log y = m.$

From (1),

$$xy = a^n . a^m = a^{n+m},$$
$$x/y = a^n \div a^m = a^{n-m},$$
$$x^k = (a^n)^k = a^{nk}.$$

Hence

$$\log(xy) = n+m = \log x+\log y,$$
$$\log(x/y) = n-m = \log x-\log y,$$
$$\log(x^k) = nk \quad = k\log x.$$

Let x be any positive number, and let $\log_e x = y$; so that $x = e^y$. For convenience of writing, let us take the suffix e to be understood as the base of all logarithms that occur hereafter. With this convention,

$$\log x = y, \qquad x = e^y. \tag{1}$$

By the fundamental property used in § 1.1,

$$\frac{d}{dy}(e^y) = e^y.$$

Hence

$$\frac{d}{dx}(x) = \frac{d}{dx}(e^y) = e^y\frac{dy}{dx};$$

that is,

$$1 = x\frac{dy}{dx}, \qquad \text{and so} \quad \frac{dy}{dx} = \frac{1}{x}.$$

Thus

$$\frac{d}{dx}(\log x) = \frac{1}{x}. \tag{2}$$

Hence
$$\int \frac{dx}{x} = \log x + C,$$
and, since $e^0 = 1$ and so $\log 1 = 0$,
$$\int_1^t \frac{dx}{x} = \log t - \log 1 = \log t. \tag{3}$$

The equations (2) and (3) embody two important properties of logarithms to the base e. In fact, one method of developing the theory takes (3) as the *definition* of $\log t$.

2.2. *The series representing* $\log(1+x)$. We first change the form of (3) by means of the substitutions
$$x = 1+y, \qquad t = 1+z.$$

When $x = 1$, $y = 0$ and when $x = t$, $y = z$; thus (3) becomes
$$\int_0^z \frac{dy}{1+y} = \log(1+z).$$

When z lies between -1 and $+1$, the values of y in the range of integration also lie in this range and
$$\frac{1}{1+y} = 1-y+y^2-y^3+\dots.$$

Thus
$$\log(1+z) = \int_0^z (1-y+y^2-y^3+\dots)\,dy.$$

Assuming (what can in fact be proved by the theory of infinite series) that the integral is given correctly by integrating each term separately,† we have
$$\log(1+z) = z-\frac{z^2}{2}+\frac{z^3}{3}-\dots.$$

Thus (changing the variable from z to x), when $-1 < x < 1$,
$$\log(1+x) = x-\frac{x^2}{2}+\frac{x^3}{3}-\dots \tag{4}$$

† A method that avoids the use of infinite series is given in W. L. Ferrar, *A Text-book of Convergence* (Oxford, 1938), p. 103. It is elementary in scope, but is rather 'tricky' for the beginner.

Further,† THE RESULT REMAINS TRUE WHEN $x = +1$, so that (4) is true when $-1 < x \leqslant 1$. It is not true when $x = -1$. As x approaches the value -1, $1+x$ approaches 0; $\log(1+x)$ is then negative and its numerical magnitude increases beyond all bounds (it tends to 'minus infinity').

To sum up,

$$\log(1+x) = x - \frac{x^2}{2} + \frac{x^3}{3} - \dots \quad (-1 < x \leqslant 1), \tag{5}$$

the *general term* of the series being

$$(-1)^{n+1}\frac{x^n}{n}.$$

This series is called the LOGARITHMIC SERIES.

2.3. As with the expansions of other functions, we must be able to recognize the series when x is replaced by simple functions of x, such as $-x$ or x^2. We note especially

$$\begin{aligned}\log(1-x) &= -x - \frac{x^2}{2} - \frac{x^3}{3} - \dots \\ &= -\left(x + \frac{x^2}{2} + \frac{x^3}{3} + \dots\right),\end{aligned}$$

which is valid when $-1 \leqslant x < 1$.

Further, any two such series can be added or subtracted; for example,

$$\log(1+x) = x - \frac{x^2}{2} + \frac{x^3}{3} - \dots,$$

$$\log(1-x) = -x - \frac{x^2}{2} - \frac{x^3}{3} - \dots,$$

and so

$$\begin{aligned}\log\frac{1+x}{1-x} &= \log(1+x) - \log(1-x) \\ &= 2\left(x + \frac{x^3}{3} + \frac{x^5}{5} + \dots\right).\end{aligned}$$

2.4. *Expansion of* $\log(2+x)$ *and similar expansions.* In using (5) we must remember that it is $1+x$ that is in question. To

† The proof of this depends on a theorem about power series that lies outside the scope of this book. The result is proved (in a different way) in *A Text-book of Convergence*, loc. cit.

deal with $2+x$ we write it as $2(1+\frac{1}{2}x)$. Thus

$$\begin{aligned}\log(2+x) &= \log\{2(1+\tfrac{1}{2}x)\}\\ &= \log 2+\log(1+\tfrac{1}{2}x)\\ &= \log 2+\frac{1}{2}x-\frac{\frac{1}{4}x^2}{2}+\frac{\frac{1}{8}x^3}{3}-\dots,\end{aligned}$$

the coefficient of x^n being $(-1)^{n+1}/n.2^n$.

2.5. *Worked examples.*

EXAMPLE 1. *Expand* $\log(1-3x+2x^2)$ *in a series of ascending powers of* x. *What is the general term of the series and for what values of* x *is the expansion valid?*

SOLUTION. $\quad 1-3x+2x^2 = (1-x)(1-2x).$

$$\therefore\ \log(1-3x+2x^2) = \log(1-x)+\log(1-2x).$$

$$\log(1-x) = -\left(x+\frac{x^2}{2}+\frac{x^3}{3}+\dots+\frac{x^n}{n}+\dots\right).$$

$$\log(1-2x) = -\left(2x+\frac{4x^2}{2}+\frac{8x^3}{3}+\dots+\frac{2^nx^n}{n}+\dots\right).$$

$$\therefore\ \log(1-3x+2x^2) = -\left(3x+\frac{5x^2}{2}+3x^3+\dots+\frac{2^n+1}{n}x^n+\dots\right).$$

The series for $\log(1-x)$ is valid when $-1 \leqslant x < 1$ and the series for $\log(1-2x)$ is valid when $-1 \leqslant 2x < 1$, i.e. when $-\frac{1}{2} \leqslant x < \frac{1}{2}$. Hence the series obtained is valid when

$$-\tfrac{1}{2} \leqslant x < \tfrac{1}{2}.$$

EXAMPLE 2. *Find the expansion of* $\log(1+2x+3x^2)$ *correct to terms in* x^4.

SOLUTION. [The quadratic $1+2x+3x^2$ has no real factors and we cannot follow the method of Example 1.]

Use the expansion

$$\log(1+y) = y-\tfrac{1}{2}y^2+\tfrac{1}{3}y^3-\tfrac{1}{4}y^4+\dots$$

with $y = 2x+3x^2 = x(2+3x)$. We get

$$\begin{aligned}\log(1+2x+3x^2) &\simeq x(2+3x)-\tfrac{1}{2}x^2(2+3x)^2+\\ &\qquad +\tfrac{1}{3}x^3(2+3x)^3-\tfrac{1}{4}x^4(2+3x)^4\\ &\simeq 2x+3x^2-\tfrac{1}{2}x^2(4+12x+9x^2)+\\ &\qquad +\tfrac{1}{3}x^3(8+36x)-\tfrac{1}{4}x^4.16,\end{aligned}$$

on neglecting all terms that would involve powers of x higher than the fourth. This gives

$$\log(1+2x+3x^2) \simeq 2x+x^2-\tfrac{10}{3}x^3+\tfrac{7}{2}x^4.$$

EXAMPLE 3. *Expand*

$$(1-x)\log(1-x)$$

in a series of ascending powers of x, giving the first three terms and the term involving x^n.

SOLUTION.

$$\log(1-x) = -(x+\tfrac{1}{2}x^2+\tfrac{1}{3}x^3+\ldots),$$

$$x\log(1-x) = -(x^2+\tfrac{1}{2}x^3+\ldots).$$

$$\therefore\quad (1-x)\log(1-x) = -x+(1-\tfrac{1}{2})x^2+(\tfrac{1}{2}-\tfrac{1}{3})x^3+\ldots$$

$$= -x+\tfrac{1}{2}x^2+\tfrac{1}{6}x^3+\ldots.$$

The coefficient of x^n in this expansion is

$$\frac{1}{n-1}-\frac{1}{n} = \frac{n-(n-1)}{n(n-1)} = \frac{1}{n(n-1)},$$

and the term involving x^n is $x^n/n(n-1)$.

EXAMPLES XI A

1. Find the first 4 terms in the expansion of

$$\log\frac{1+x}{(1-x)^2}$$

in a series of ascending powers of x; and state the range of values of x for which the expansion is valid.

2. Write down the expansions of e^x, e^{-2x}, and e^{-3x}. Prove that, correct to terms in x^3,

$$e^x+2e^{-2x}-e^{-3x} \simeq 2(1+x^3).$$

3. Prove that $\quad \dfrac{e^x+e^{-x}}{e^x-e^{-x}} = \dfrac{e^{2x}+1}{e^{2x}-1} = \dfrac{1+e^{-2x}}{1-e^{-2x}}$

4. Prove that in the expansion of $(1+x)e^x$ as a power series in x the coefficient of x^n is $(n+1)/(n!)$.

5. Find the coefficient of x^n in the expansion of

$$(1+x)e^x-(1-x)e^{-x}$$

(i) when n is even, (ii) when n is odd. You may use the result of Example 4.

6. Prove that, when $-1 < x < 1$,

$$\frac{1}{2}\log\frac{1+x}{1-x} = x+\frac{1}{3}x^3+\frac{1}{5}x^5+\dots,$$

and deduce that, when $u > 1$,

$$\log(\sqrt{u}) = \frac{u-1}{u+1}+\frac{1}{3}\left(\frac{u-1}{u+1}\right)^3+\frac{1}{5}\left(\frac{u-1}{u+1}\right)^5+\dots.$$

Prove that $\log(\frac{1}{3}\sqrt{11}) \simeq 0{\cdot}100335$.

7. Write down the series for e^{-2x} and e^{-4x}, and deduce that

$$\frac{e^x+e^{-x}}{e^{3x}} = 2-6x+20\frac{x^2}{2!}+\dots+(-1)^n\frac{(4^n+2^n)x^n}{n!}+\dots.$$

8. The function

$$e^{-x}-(1-x)(1-x^2)^{-\frac{1}{2}}(1-x^3)^{-\frac{1}{3}}$$

is expanded in ascending powers of x in the form

$$a+bx+cx^2+dx^3+ex^4+\dots.$$

Prove that the first 5 coefficients a, b, c, d, e are all zero.

9. (i) Expand $\log(6-5x)-\log 6$ in a series of ascending powers of x. For what values of x is the expansion valid?

(ii) Find the coefficient of x^n when $\log\{1/(6-5x+x^2)\}$ is expanded in a series of ascending powers of x. For what values of x is the expansion valid?

10. Expand $$\log\frac{2+x}{1-x}$$

in ascending powers of x, giving the first 3 terms and the general term of the expansion. For what values of x is the expansion valid?

11. Expand $\log(1+4x+5x^2)$ and $\log(1+5x+6x^2)$ correct to terms in x^3. Hence prove that

$$\log\frac{1+4x+5x^2}{1+5x+6x^2} \simeq -x+\frac{7}{2}x^2-\frac{31}{3}x^3$$

correct to terms in x^3.

12. Find the first 4 terms in the expansion of $\log\{(1-x^3)/(1-x)\}$, and prove that the coefficient of x^{3n} in this expansion is $-2/(3n)$.

EXAMPLES XI B. HARDER EXAMPLES

1. Determine a and b so that the expansion of

$$\frac{1+ax}{1+bx}\log(1+x)$$

may contain no term in x^2 or x^3, and show that with these values of a and b

$$\frac{1+bx}{1+ax} \simeq 1-\frac{x}{2}+\frac{x^2}{3}-\frac{2x^3}{9},$$

correct to terms in x^3.

2. Show that, if $x = 1+at$ where t is small,

$$e^x - e^{-x} \simeq \left(e - \frac{1}{e}\right) + at\left(e + \frac{1}{e}\right) + \tfrac{1}{2}a^2t^2\left(e - \frac{1}{e}\right)$$

correct to terms in t^2.

3. What logarithmic function is represented by the series

$$x + \frac{x^3}{3} + \frac{x^5}{5} + \dots ?$$

Prove that, if this series represents $f(x)$, then

$$f\left(\frac{2x}{1+x^2}\right) = 2f(x).$$

4. What function is represented by the series

$$1 + \frac{x^2}{3} + \frac{x^4}{5} + \dots ?$$

Prove that, if this series represents $F(x)$,

$$F\left(\frac{2x}{1+x^2}\right) = (1+x^2)F(x).$$

CHAPTER XII*

MATHEMATICAL INDUCTION

1. Introduction

1.1. In this chapter we explain a method that is useful in many parts of algebra. This is the method of 'proof by induction'. At first the method furnished both an instrument of research to discover new theorems and a proof of the theorems discovered: it also provided relatively simple proofs of theorems discovered by other means. The reader will probably use the method only as a means of proving results he is asked to prove and will rarely, if ever, be called upon to find out for himself just what the theorem is before he tries to prove it. But it is worth while to have some knowledge of how the method can be used to discover results and, in this introductory section, we give some illustrations of the process.

1.2. *A sum of cubes.*

PROBLEM. *To find the sum of the first n terms of the series*

$$1^3+2^3+3^3+\dots.$$

SOLUTION AND EXPLANATION.

The sum of 1 term is $1^3 = 1$,
the sum of 2 terms is $1^3+2^3 = 9$,
the sum of 3 terms is $1^3+2^3+3^3 = 36$.

If we look for any particular property of the numbers 1, 9, 36, we see at once that they are squares; in fact

the sum of 1 term is 1^2,
the sum of 2 terms is 3^2,
the sum of 3 terms is 6^2,
the sum of 4 terms is 10^2.

Let us write down the numbers

$$1,\ 3,\ 6,\ 10,\ \dots. \qquad (1)$$

As they stand they suggest no immediate connexion with the numbers

$$1,\ 2,\ 3,\ 4,\ \dots, \qquad (2)$$

but if we multiply the numbers in (1) by 2, we get

$$2, 6, 12, 20, \ldots. \tag{3}$$

We can now see that each number in (2) is a factor of the corresponding number in (3), and we write (3) as

$$1.2, \quad 2.3, \quad 3.4, \quad 4.5, \quad \ldots. \tag{4}$$

We have thus found out that

$$1^3 = (\tfrac{1}{2}.1.2)^2,$$
$$1^3+2^3 = (\tfrac{1}{2}.2.3)^2,$$
$$1^3+2^3+3^3 = (\tfrac{1}{2}.3.4)^2,$$
$$1^3+2^3+3^3+4^3 = (\tfrac{1}{2}.4.5)^2.$$

It is suggested, but not of course yet *proved*, that for every positive integer n

$$1^3+2^3+\ldots+n^3 = \{\tfrac{1}{2}n(n+1)\}^2. \tag{5}$$

To complete this proof we may proceed as follows:

Let us suppose that, for some definite value of r,

$$1^3+2^3+\ldots+r^3 = \{\tfrac{1}{2}r(r+1)\}^2,$$

a supposition we know to be true when $r = 1$, 2, 3, or 4.

Then, on this assumption,

$$\begin{aligned} 1^3+2^3+\ldots+r^3+(r+1)^3 &= \tfrac{1}{4}r^2(r+1)^2+(r+1)^3 \\ &= \tfrac{1}{4}(r+1)^2\{r^2+4(r+1)\} \\ &= \tfrac{1}{4}(r+1)^2(r+2)^2 \end{aligned}$$

which is the R.H.S. of (5) when $n = r+1$. Thus, if (5) is true when $n = r$, it is also true when $n = r+1$. But we know that (5) is true when $n = 4$; it is therefore true when $n = 5$. Again, since (5) is true when $n = 5$, it is also true when $n = 6$; and we may proceed thus to show that it is true when n is any positive integer whatsoever.

1.3. NOTE. The above is an introductory example, showing how we may, from a study of the results for the first few values of n, guess the answer and then go on to prove it. The foregoing work is not a model for doing later examples: see § 3.

1.4. PROBLEM. *What numerical factor, if any, is common to all numbers*

$$6^n-5n+4$$

when n is a positive integer?

SOLUTION.

Denote 6^n-5n+4 by $f(n)$. Then

$$f(1) = 6-5+4 = 5,$$
$$f(2) = 36-10+4 = 30,$$

and it is suggested that 5 is always a factor of $f(n)$.

Suppose then that, for a particular value of n, $f(n) = 5m$, where m is an integer. Then

$$f(n+1)-f(n) = 6^{n+1}-6^n-5(n+1)+5n$$
$$= 6^n(6-1)-5 = 5(6^n-1).$$

Hence $$f(n+1) = 5m+5(6^n-1),$$

which contains 5 as a factor.

Thus, if $f(n)$ is divisible by 5 when n has any particular value, $f(n+1)$ is also divisible by 5. But $f(1)$ and $f(2)$ are divisible by 5, and so $f(3)$ is divisible. Since $f(3)$ is divisible by 5, so also is $f(4)$; and so on for all positive integers.

2. Formal statement

THE PRINCIPLE OF MATHEMATICAL INDUCTION.

A mathematical formula involving the positive integer n is true for ALL *positive integers provided that*

(1) *it is true when $n = 1$,*

and (2) *the hypothesis that it is true for any particular n is sufficient to ensure that it is also true for $n+1$.*

The principle is established thus: Suppose that a formula satisfies the conditions (1) and (2). Then, by (2), if the formula is true for n, it is also true for $n+1$; so that,

since, by (1), it is true when $n = 1$, it is also true when $n = 2$;
since it is true when $n = 2$, it is also true when $n = 3$;
since it is true when $n = 3$, it is also true when $n = 4$;
and so on for all values of n.

3. Worked examples

EXAMPLE 1. *Prove by induction that*

$$1^3+2^3+\ldots+n^3 = \tfrac{1}{4}n^2(n+1)^2.$$

PROOF. (1) The formula is true when $n = 1$, since

$$1^3 = \tfrac{1}{4}.1^2.2^2.$$

(2) Assume that, *for a particular value of* n,

$$1^3+2^3+...+n^3 = \tfrac{1}{4}n^2(n+1)^2. \qquad \text{(A)}$$

Then

$$\begin{aligned} 1^3+2^3+...+n^3+(n+1)^3 &= \tfrac{1}{4}n^2(n+1)^2+(n+1)^3 \\ &= \tfrac{1}{4}(n+1)^2\{n^2+4(n+1)\} \\ &= \tfrac{1}{4}(n+1)^2(n+2)^2, \end{aligned}$$

that is $\quad 1^3+2^3+...+(n+1)^3 = \tfrac{1}{4}(n+1)^2(n+2)^2.$

This is the original formula (A) with $n+1$ instead of n. Hence, if (A) is true for n, it is also true for $n+1$.

But we have shown in (1) that the formula is true when $n = 1$, and therefore, by the principle of induction, it is true for ALL positive integers n.

EXAMPLE 2. *Prove by induction that, if n is a positive integer,*

$$3^{2n+2}-8n-9$$

is divisible by 64.

PROOF.

(1) Let $f(n)$ denote $3^{2n+2}-8n-9$. Then

$$f(1) = 3^4-17 = 64.$$

(2)
$$\begin{aligned} f(n+1)-f(n) &= 3^{2n+4}-8(n+1)-3^{2n+2}+8n \\ &= 3^{2n+2}(3^2-1)-8 \\ &= 8(3^{2n+2}-1) \\ &= 8\{(3^2)^{n+1}-1\}. \end{aligned}$$

Also, $\quad x^{n+1}-1 \equiv (x-1)(x^n+x^{n-1}+...+1),$

so that $\quad 9^{n+1}-1 = 8(9^n+9^{n-1}+...+1).$

Hence $f(n+1)-f(n)$ is divisible by 64, and therefore, if $f(n)$ is divisible by 64, so also is $f(n+1)$.

But we have shown in (1) that $f(1)$ is divisible by 64, and therefore, by the principle of induction, $f(n)$ is divisible by 64 for all positive integers n.

EXAMPLE 3. *Prove by induction that, if n is a positive integer,*

$$1^2+2^2+...+n^2 = \tfrac{1}{6}n(n+1)(2n+1).$$

PROOF. We shall set out the proof of this result in a way

that differs in its detail from the way used to prove Example 1. The reader may take his choice of the two ways when doing examples for himself.

(1) The result is true when $n = 1$, since

$$1^2 = \tfrac{1}{6}.1.2.3.$$

(2) Let $f(n)$ denote $\tfrac{1}{6}n(n+1)(2n+1)$. Then

$$\begin{aligned} f(n+1)-f(n) &= \tfrac{1}{6}(n+1)(n+2)(2n+3)-\tfrac{1}{6}n(n+1)(2n+1) \\ &= \tfrac{1}{6}(n+1)\{2n^2+7n+6-2n^2-n\} \\ &= \tfrac{1}{6}(n+1)(6n+6) \\ &= (n+1)^2. \end{aligned} \qquad \text{(A)}$$

Hence, if the formula

$$1^2+2^2+...+n^2 = \tfrac{1}{6}n(n+1)(2n+1) = f(n)$$

is true for a particular value of n, it follows that

$$\begin{aligned} 1^2+2^2+...+(n+1)^2 &= f(n)+(n+1)^2 \\ &= f(n+1), \end{aligned}$$

by (A); i.e. the formula is also true for the value $n+1$.

But we have shown in (1) that the formula is true when $n = 1$, and therefore, by the principle of induction, it is true for all positive integers n.

4. Proof by induction of the binomial theorem

The binomial theorem (cf. Theorem 16) states that, when n is a positive integer,

$$(a+x)^n = a^n+na^{n-1}x+...+{}_nC_r\,a^{n-r}x^r+...+nax^{n-1}+x^n. \qquad \text{(A)}$$

The theorem may be proved by induction thus:

(1) When $n = 1$, the R.H.S. of (A) reduces simply to $a+x$, and so the formula (A) is true when $n = 1$.

(2) Assume that (A) is true for a particular value of n. Then, on multiplying (A) by $a+x$,†

$$\begin{aligned} (a+x)^{n+1} &= a^{n+1}+na^nx+...+{}_nC_r\,a^{n-r+1}x^r+...+ax^n+ \\ &\qquad +a^nx+...+{}_nC_{r-1}\,a^{n-r+1}x^r+...+nax^n+x^{n+1} \\ &= a^{n+1}+(n+1)a^nx+...+({}_nC_r+{}_nC_{r-1})a^{n-r+1}x^r+ \\ &\qquad +...+x^{n+1}. \end{aligned}$$

† The first line gives the terms of (A) multiplied by a, the second line gives the terms of (A) multiplied by x.

But
$$ {}_nC_r + {}_nC_{r-1} = \frac{n!}{r!\,(n-r)!} + \frac{n!}{(r-1)!\,(n-r+1)!} $$
$$ = \frac{n!}{r!\,(n-r+1)!}(n-r+1+r) $$
$$ = \frac{(n+1)!}{r!\,(n+1-r)!} $$
$$ = {}_{n+1}C_r. $$

Hence, if (A) is true for a particular value of n, it follows that
$$ (a+x)^{n+1} = a^{n+1} + (n+1)a^n x + \ldots + {}_{n+1}C_r\, a^{n+1-r} x^r + \ldots + x^{n+1}, $$
that is, (A) is also true for the value $n+1$.

But we have shown in (1) that the formula (A) is true when $n = 1$, and so, by the principle of induction, it is true for all positive integers.

EXAMPLES XII

Prove by induction the results in the following examples:

1. $\dfrac{1}{1.2} + \dfrac{1}{2.3} + \ldots + \dfrac{1}{n(n+1)} = \dfrac{n}{n+1}.$

2. $1.2 + 2.3 + \ldots + n(n+1) = \frac{1}{3}n(n+1)(n+2).$

3. $1+3+5+\ldots+(2n-1) = n^2.$

4. (*Harder.*)
$$ 1^5 + 2^5 + \ldots + n^5 = \tfrac{1}{12}n^2(n+1)^2(2n^2+2n-1). $$

5. $7^{2n} - 48n - 1$ is divisible by 2304 if n is a positive integer greater than unity.

HINT. The method of induction applies, though here the result (1), cf. the worked examples, concerns $n = 2$ and the induction applies to all $n \geqslant 2$. When $n = 1$, the given expression is zero.

6. (*Harder.*) $9^n - 8n - 1$ is divisible by 64 if n is a positive integer greater than unity.

HINT. $3^n - 1$ and $3^n + 1$ are consecutive even integers; or
$$ (3^n - 1)(3^n + 1) = 9^n - 1, $$
which contains 8 as a factor.

7. (*Harder.*) $1^2 + 4^2 + 7^2 + \ldots + (3n-2)^2 = \frac{1}{2}n(6n^2 - 3n - 1).$

8. $1^2 + 3^2 + 5^2 + \ldots + (2n-1)^2 = \frac{1}{3}n(4n^2 - 1).$

9. (*Harder.*) Prove that, if $p_0 = 1$ and $p_1, p_2, \ldots$ are positive numbers such that $p_{r+1}^2 > p_r p_{r+2}$ $(r = 0, 1, 2, \ldots)$, then $p_1 > p_2^{\frac{1}{2}} > p_3^{\frac{1}{3}} \ldots > p_n^{1/n}$.

CHAPTER XIII

LINEAR EQUATIONS AND DETERMINANTS

1. Simultaneous equations

1.1. *Introduction.* The two equations

$$\left.\begin{aligned} 2x+y &= 3 \\ 5x-4y &= 1 \end{aligned}\right\} \qquad \text{(I)}$$

can be solved for x and y; the solution is $x = 1$, $y = 1$, and no other pair of values will satisfy both equations.

The two equations

$$\left.\begin{aligned} x+y-z &= 0 \\ 5x+3y-4z &= 0 \end{aligned}\right\} \qquad \text{(II)}$$

do not determine the actual values of x, y, z, but only their ratios. The two equations may be written as

$$\frac{x}{z}+\frac{y}{z} = 1, \qquad 5\frac{x}{z}+3\frac{y}{z} = 4,$$

and when we solve these equations for x/z and y/z we get

$$\frac{x}{z} = \frac{1}{2}, \qquad \frac{y}{z} = \frac{1}{2},$$

i.e. $$x:y:z = 1:1:2.$$

If λ is any number whatsoever, the values

$$x = \lambda, \qquad y = \lambda, \qquad z = 2\lambda$$

satisfy the equations (II).

The equations (I) are said to be non-homogeneous in the two variables x and y, since they contain numbers, namely 3 and 1, on the right-hand side; they have one solution and one only.

The equations (II) are said to be homogeneous in the three variables x, y, z, since each term is of the first degree and the numbers on the right-hand side are both zeros. The equations are satisfied when

$$\frac{x}{1} = \frac{y}{1} = \frac{z}{2},$$

i.e. when $x = \lambda$, $y = \lambda$, $z = 2\lambda$ and λ is any number whatsoever.

In § 1.2 we consider the same type of equation with literal coefficients.

1.2. *Solution of two simultaneous equations.* Consider the simultaneous equations

$$a_1x+b_1y+c_1z = 0, \tag{1}$$

$$a_2x+b_2y+c_2z = 0. \tag{2}$$

Multiply (1) by b_2, (2) by b_1, and subtract; this gives

$$(a_1b_2-a_2b_1)x = (b_1c_2-b_2c_1)z. \tag{3}$$

Multiply (2) by a_1, (1) by a_2, and subtract; this gives

$$(a_1b_2-a_2b_1)y = -(a_1c_2-a_2c_1)z. \tag{4}$$

Provided that no one of the expressions

$$a_1b_2-a_2b_1, \quad b_1c_2-b_2c_1, \quad a_1c_2-a_2c_1$$

is zero,† we may combine (3) and (4) to give the formula

$$\frac{x}{b_1c_2-b_2c_1} = \frac{-y}{a_1c_2-a_2c_1} = \frac{z}{a_1b_2-a_2b_1}. \tag{5}$$

If λ is any number whatsoever, the values

$$x = \lambda(b_1c_2-b_2c_1), \qquad y = -\lambda(a_1c_2-a_2c_1),$$
$$z = \lambda(a_1b_2-a_2b_1) \tag{6}$$

satisfy (5). We now verify that these values satisfy the original equations (1) and (2).

In the first place,

$$\lambda[a_1(b_1c_2-b_2c_1)-b_1(a_1c_2-a_2c_1)+c_1(a_1b_2-a_2b_1)] = 0, \tag{7}$$

the terms cancelling in pairs, and so the values (6) satisfy equation (1); also

$$\lambda[a_2(b_1c_2-b_2c_1)-b_2(a_1c_2-a_2c_1)+c_2(a_1b_2-a_2b_1)] = 0, \tag{8}$$

and so the values (6) satisfy equation (2).

NOTE. It is often useful to be able to write down (5) or, what is the same thing, (6) as the solution of the equations (1) and (2). Notice carefully the way in which the denominators of (5)

† If the reader meets examples in which one of the denominators in (5) is zero, he should, for the present, put aside the formula (5) and solve the equations *ab initio*; he will then obtain equations like (3) and (4), but with one side zero. At a later stage he may learn how to interpret the formula (5) when a denominator is zero.

are obtained by 'cross-multiplication' on the pattern

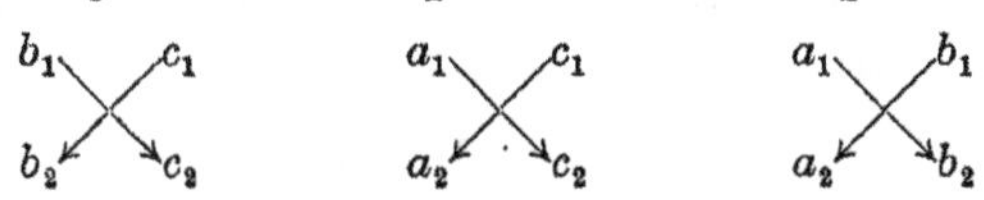

and the arrangement of the signs† of the numerators.

2. Determinants of order two

The expressions (7) and (8) are examples of what are called 'determinants'. Such expressions occur in all attempts to solve linear equations and a special notation has been invented to enable us to deal with them. The expressions (7) and (8) are built up from the simpler expressions $b_1 c_2 - b_2 c_1$, $a_1 c_2 - a_2 c_1$, $a_1 b_2 - a_2 b_1$ that occur in (5). In turn, expressions resembling (7) and (8) can be used to build up expressions in four letters a, b, c, d and so on for expressions in five, six,... letters. We begin by considering the very simplest type of determinant.

The notation
$$\begin{vmatrix} b_1 & c_1 \\ b_2 & c_2 \end{vmatrix},$$
called a determinant, is used to denote $b_1 c_2 - b_2 c_1$, and so for other letters and suffixes. For example,

$$\begin{vmatrix} x & y \\ u & v \end{vmatrix} = xv - yu, \qquad \begin{vmatrix} l & m \\ 1 & 1 \end{vmatrix} = l - m,$$

$$\begin{vmatrix} 4 & -1 \\ 5 & 7 \end{vmatrix} = 28+5 = 33, \qquad \begin{vmatrix} -1 & 4 \\ -3 & -1 \end{vmatrix} = 1+12 = 13.$$

3. Properties of determinants of order two

We note the following facts and properties chiefly as a preparation for theorems concerning determinants of order three, for which similar facts and properties hold, but are not quite

† Some writers use the cyclic order
$$b_1 c_2 - b_2 c_1, \quad c_1 a_2 - c_2 a_1, \quad a_1 b_2 - a_2 b_1,$$
that is, they write $c_1 a_2 - c_2 a_1$ where we have written $-(a_1 c_2 - a_2 c_1)$, and have all the signs positive in the numerators. This is right, of course, but it often leads to error when the student comes to deal with determinants of order four; determinants are founded on the alternating signs, $+, -, +$, as we have them in (5), and 'cyclic order' has the grave fault that it sometimes leads to correct results [as it does in (5)] and sometimes to wrong results, as it does with four letters a, b, c, d.

so obvious and elementary. For the purposes of this section we take as the standard determinant of order two

$$\begin{vmatrix} a_1 & b_1 \\ a_2 & b_2 \end{vmatrix}.$$

I. The determinant has two rows; a first row a_1, b_1 and a second row a_2, b_2. It has two columns; a first column a_1, a_2 and a second column b_1, b_2. The individual letters, a_1, a_2, b_1, b_2 are sometimes called 'ELEMENTS'.

II. The value of each of the determinants

$$\begin{vmatrix} a_1 & b_1 \\ a_2 & b_2 \end{vmatrix}, \qquad \begin{vmatrix} a_1 & a_2 \\ b_1 & b_2 \end{vmatrix}$$

is $a_1 b_2 - a_2 b_1$; that is, *the value of the determinant is unaltered if rows become columns and columns become rows.*

III. *The interchange of two columns, or of two rows, in the determinant multiplies the value of the determinant by* -1. For example,

$$\begin{vmatrix} a_1 & b_1 \\ a_2 & b_2 \end{vmatrix} = -\begin{vmatrix} b_1 & a_1 \\ b_2 & a_2 \end{vmatrix}.$$

PROOF. The L.H.S. $= a_1 b_2 - a_2 b_1$

and the R.H.S. $= -(a_2 b_1 - a_1 b_2) = a_1 b_2 - a_2 b_1$.

IV. *If a determinant has two columns, or two rows, identical its value is zero.*

PROOF. The determinant

$$\begin{vmatrix} a & a \\ b & b \end{vmatrix}$$

has its second column the same as its first. Its value is

$$ab - ab = 0.$$

V. *If each element of one column, or row, is multiplied by a factor K, the value of the determinant is thereby multiplied by K.*

PROOF.
$$\begin{vmatrix} Ka_1 & b_1 \\ Ka_2 & b_2 \end{vmatrix} = Ka_1 b_2 - Ka_2 b_1$$
$$= K(a_1 b_2 - a_2 b_1)$$
$$= K \times \begin{vmatrix} a_1 & b_1 \\ a_2 & b_2 \end{vmatrix},$$

and so for the second column or either row.

COROLLARY TO IV AND V. *The value of the determinant*

$$\begin{vmatrix} Ka_1 & a_1 \\ Ka_2 & a_2 \end{vmatrix},$$

in which the elements of one column are K times the corresponding elements of the other column, is zero; for, by V, it is K times the determinant

$$\begin{vmatrix} a_1 & a_1 \\ a_2 & a_2 \end{vmatrix},$$

which, by IV, is equal to zero.

VI. *The value of a determinant is unaltered if to each element of one column (or row) is added the same multiple of the corresponding element of another column (or row); for example,*

$$\begin{vmatrix} a_1 & b_1 \\ a_2 & b_2 \end{vmatrix} = \begin{vmatrix} a_1+kb_1 & b_1 \\ a_2+kb_2 & b_2 \end{vmatrix},$$

$$\begin{vmatrix} a_1 & b_1 \\ a_2 & b_2 \end{vmatrix} = \begin{vmatrix} a_1 & b_1 \\ a_2+5a_1 & b_2+5b_1 \end{vmatrix},$$

and so on.

PROOF. We shall prove that

$$\begin{vmatrix} a_1 & b_1 \\ a_2 & b_2 \end{vmatrix} = \begin{vmatrix} a_1+kb_1 & b_1 \\ a_2+kb_2 & b_2 \end{vmatrix}$$

for all values of k. The value of the second determinant is

$$(a_1+kb_1)b_2-(a_2+kb_2)b_1 = a_1b_2-a_2b_1+k(b_1b_2-b_2b_1)$$

$$= a_1b_2-a_2b_1 = \begin{vmatrix} a_1 & b_1 \\ a_2 & b_2 \end{vmatrix}.$$

The proof for other additions is similar.

NOTATION FOR APPLICATIONS OF VI. When we apply the result VI to examples we use r_1, r_2 to denote the first and second rows of the original determinant, and we use r_1', r_2' to denote the first and second rows of the new determinant obtained by applying VI. To indicate that the second row (say) of the new determinant is the original second row plus five times the original first row we write

$$r_2' = r_2+5r_1,$$

and so for multiples other than 5. For example, we write

$$\begin{vmatrix} -30 & 19 \\ 91 & -57 \end{vmatrix} = \begin{vmatrix} -30 & 19 \\ 1 & 0 \end{vmatrix} \qquad (r_2' = r_2+3r_1),$$

the notation $r_2' = r_2+3r_1$ indicating that we get the second row of the determinant on the right by adding to the original second row three times the first row; or again,

$$\begin{vmatrix} -15 & 70 \\ 10 & -62 \end{vmatrix} = \begin{vmatrix} -5 & 8 \\ 10 & -62 \end{vmatrix} \qquad (r_1' = r_1+r_2),$$

the notation $r_1' = r_1+r_2$ indicating that we get the first row of the determinant on the right by adding the two rows of the determinant on the left.

When dealing with columns we use the letter c instead of r, which indicates rows. For example, we write

$$\begin{vmatrix} a+x & x \\ b+y & y \end{vmatrix} = \begin{vmatrix} a & x \\ b & y \end{vmatrix} \qquad (c_1' = c_1-c_2);$$

or again

$$\begin{vmatrix} a & 3a+4b \\ b & 5a+3b \end{vmatrix} = \begin{vmatrix} a & 4b \\ b & 5a \end{vmatrix} \qquad (c_2' = c_2-3c_1).$$

EXAMPLES XIII A

Find the values of the following determinants:

1. $\begin{vmatrix} 7 & 5 \\ 9 & 8 \end{vmatrix}, \quad \begin{vmatrix} 7 & -5 \\ 9 & 3 \end{vmatrix}, \quad \begin{vmatrix} 1 & 2 \\ 3 & 4 \end{vmatrix}, \quad \begin{vmatrix} 2 & -3 \\ 19 & -7 \end{vmatrix}.$

2. $\begin{vmatrix} x^2 & y^2 \\ x & y \end{vmatrix}, \quad \begin{vmatrix} a+x & a \\ b+y & b \end{vmatrix}, \quad \begin{vmatrix} x+1 & x^2+1 \\ y+1 & y^2+1 \end{vmatrix}.$

3. $\begin{vmatrix} 4 & -2 \\ 7 & 9 \end{vmatrix}, \quad \begin{vmatrix} -1 & -5 \\ 9 & -6 \end{vmatrix}, \quad \begin{vmatrix} 23 & 3 \\ -2 & 2 \end{vmatrix}, \quad \begin{vmatrix} 7 & -3 \\ -1 & 8 \end{vmatrix}.$

Prove the following results by using properties II–VI of § 3.

4. $\begin{vmatrix} 59 & 72 \\ 61 & -31 \end{vmatrix} = \begin{vmatrix} 59 & 61 \\ 72 & -31 \end{vmatrix}, \quad \begin{vmatrix} 19 & 42 \\ 17 & 41 \end{vmatrix} = -\begin{vmatrix} 42 & 19 \\ 41 & 17 \end{vmatrix}.$

5. $\begin{vmatrix} 19 & 19 \\ 15 & 15 \end{vmatrix} = 0, \quad \begin{vmatrix} 57 & 19 \\ 45 & 15 \end{vmatrix} = 0, \quad \begin{vmatrix} l & m \\ n & p \end{vmatrix} = \begin{vmatrix} l & n \\ m & p \end{vmatrix}.$

6. $\begin{vmatrix} 3x & u \\ 3y & v \end{vmatrix} = 3\begin{vmatrix} x & u \\ y & v \end{vmatrix}, \quad \begin{vmatrix} 57 & 18 \\ 21 & 17 \end{vmatrix} = 3\begin{vmatrix} 19 & 18 \\ 7 & 17 \end{vmatrix}.$

7. $\begin{vmatrix} x & y \\ u+5x & v+5y \end{vmatrix} = \begin{vmatrix} x & y \\ u & v \end{vmatrix}$, $\begin{vmatrix} 101 & 151 \\ 50 & 75 \end{vmatrix} = \begin{vmatrix} 1 & 1 \\ 50 & 75 \end{vmatrix}$.

Solve the following pairs of simultaneous equations, using (5) of § 1.

8. $3x+4y+5z = 0,$
$x+y+z = 0.$

9. $7x-2y+3z = 0,$
$2x+7y-4z = 0.$

10. $7x-2y+3z = 0,$
$x+y+z = 0.$

11. $3x+4y+5z = 0,$
$2x+7y-4z = 0.$

12. $15x-14y+13z = 0,$
$7x-6y+5z = 0.$

13. $17x+16y-15z = 0,$
$35x+33y-31z = 0.$

Solve the simultaneous equations obtained by putting $z = 1$ in Examples 8–13, e.g.

14. $3x+4y+5 = 0,$
$x+y+1 = 0.$

15. $7x-2y+3 = 0,$
$2x+7y-4 = 0.$

16–19 derived similarly from 10–13.

4. Determinants of order three

4.1. The notation

$$\begin{vmatrix} a_1 & b_1 & c_1 \\ a_2 & b_2 & c_2 \\ a_3 & b_3 & c_3 \end{vmatrix}, \tag{1}$$

called a determinant of order three, is used to denote

$$a_1(b_2c_3-b_3c_2)-b_1(a_2c_3-a_3c_2)+c_1(a_2b_3-a_3b_2), \tag{2}$$

which may also be written as

$$a_1\begin{vmatrix} b_2 & c_2 \\ b_3 & c_3 \end{vmatrix} - b_1\begin{vmatrix} a_2 & c_2 \\ a_3 & c_3 \end{vmatrix} + c_1\begin{vmatrix} a_2 & b_2 \\ a_3 & b_3 \end{vmatrix}. \tag{3}$$

The method of obtaining the determinants

$$\begin{vmatrix} b_2 & c_2 \\ b_3 & c_3 \end{vmatrix}, \quad \begin{vmatrix} a_2 & c_2 \\ a_3 & c_3 \end{vmatrix}, \quad \begin{vmatrix} a_2 & b_2 \\ a_3 & b_3 \end{vmatrix},$$

which occur in (3), from the original determinant (1) must be noted. They are obtained by deleting from (1) the first row in every case, and then deleting in turn from (1) the first, second, third column respectively.

4.2. *Worked examples.* Before proceeding further we give the reader some practice in writing down the value of a determinant of order three.

$$\begin{vmatrix} 5 & 6 & 7 \\ 4 & 3 & 2 \\ 7 & 1 & 2 \end{vmatrix} = 5\begin{vmatrix} 3 & 2 \\ 1 & 2 \end{vmatrix} - 6\begin{vmatrix} 4 & 2 \\ 7 & 2 \end{vmatrix} + 7\begin{vmatrix} 4 & 3 \\ 7 & 1 \end{vmatrix}$$

$$= 5(6-2)-6(8-14)+7(4-21)$$

$$= 20+36-119 = -63.$$

$$\begin{vmatrix} a & h & g \\ h & b & f \\ g & f & c \end{vmatrix} = a\begin{vmatrix} b & f \\ f & c \end{vmatrix} - h\begin{vmatrix} h & f \\ g & c \end{vmatrix} + g\begin{vmatrix} h & b \\ g & f \end{vmatrix}$$

$$= a(bc-f^2)-h(hc-fg)+g(hf-bg)$$

$$= abc+2fgh-af^2-bg^2-ch^2.$$

$$\begin{vmatrix} 1 & 1 & 1 \\ x_1 & y_1 & z_1 \\ x_2 & y_2 & z_2 \end{vmatrix} = \begin{vmatrix} y_1 & z_1 \\ y_2 & z_2 \end{vmatrix} - \begin{vmatrix} x_1 & z_1 \\ x_2 & z_2 \end{vmatrix} + \begin{vmatrix} x_1 & y_1 \\ x_2 & y_2 \end{vmatrix}$$

$$= (y_1 z_2 - y_2 z_1)+(z_1 x_2 - z_2 x_1)+(x_1 y_2 - x_2 y_1).$$

[Notice that we revert to the use of cyclic order when we write the *answer*.]

EXAMPLES XIII B

Find the values of the following determinants:

1. $\begin{vmatrix} 4 & 5 & 6 \\ 1 & 2 & 3 \\ 0 & 1 & -1 \end{vmatrix}$, $\begin{vmatrix} 4 & -1 & 2 \\ 0 & 7 & 8 \\ 1 & 2 & 3 \end{vmatrix}$, $\begin{vmatrix} 5 & 0 & 7 \\ 0 & 6 & 5 \\ 1 & -1 & 2 \end{vmatrix}$.

2. $\begin{vmatrix} x^2 & x & 1 \\ y^2 & y & 1 \\ z^2 & z & 1 \end{vmatrix}$, $\begin{vmatrix} x & y & z \\ z & x & y \\ y & z & x \end{vmatrix}$.

3. $\begin{vmatrix} ax & by & cz \\ x^2 & y^2 & z^2 \\ 1 & 1 & 1 \end{vmatrix}$, $\begin{vmatrix} a & b & c \\ x & y & z \\ yz & zx & xy \end{vmatrix}$.

4.3. *Other ways of expressing the value of a determinant.* If we collect the terms of (2) in a different way, this time collecting all terms containing a_1, then all terms containing a_2, and then all terms containing a_3, we obtain an expansion of the determinant (1) in the form

$$a_1(b_2 c_3 - b_3 c_2) - a_2(b_1 c_3 - b_3 c_1) + a_3(b_1 c_2 - b_2 c_1),$$

which may also be written as

$$a_1\begin{vmatrix} b_2 & c_2 \\ b_3 & c_3 \end{vmatrix} - a_2\begin{vmatrix} b_1 & c_1 \\ b_3 & c_3 \end{vmatrix} + a_3\begin{vmatrix} b_1 & c_1 \\ b_2 & c_2 \end{vmatrix}. \qquad (4)$$

The determinants of order two that occur in (4) are obtained from the original determinant

$$\begin{vmatrix} a_1 & b_1 & c_1 \\ a_2 & b_2 & c_2 \\ a_3 & b_3 & c_3 \end{vmatrix} \qquad (1)$$

by deleting

(i) the row and column containing a_1,

(ii) the row and column containing a_2,

(iii) the row and column containing a_3.

Now (4) works down the first column of (1), the terms of (4) being a_1 times something, minus a_2 times something, plus a_3 times something; and so (4) is called THE EXPANSION OF THE GIVEN DETERMINANT BY ITS FIRST COLUMN.

In the same way (3), which was

$$a_1\begin{vmatrix} b_2 & c_2 \\ b_3 & c_3 \end{vmatrix} - b_1\begin{vmatrix} a_2 & c_2 \\ a_3 & c_3 \end{vmatrix} + c_1\begin{vmatrix} a_2 & b_2 \\ a_3 & b_3 \end{vmatrix}, \qquad (3)$$

works from the first row a_1, b_1, c_1 of (1); and so (3) is called THE EXPANSION OF THE GIVEN DETERMINANT BY ITS FIRST ROW.

The determinant can also be expanded by its second or third row, or by its second or third column. But, for the present, the reader will be well advised to concentrate on the two expansions (3), expansion by the first row, and (4), expansion by the first column.

EXAMPLES XIII C

1. Expand the following determinants by their first rows (do not simplify the expansion by removing brackets) on the pattern of the example:

$$\begin{vmatrix} 1 & 2 & 3 \\ a & b & c \\ l & m & n \end{vmatrix} = (bn-cm)-2(an-cl)+3(am-bl):$$

$$\begin{vmatrix} x & y & z \\ l & m & n \\ a & b & c \end{vmatrix}, \quad \begin{vmatrix} x & y & z \\ a & b & c \\ l & m & n \end{vmatrix}, \quad \begin{vmatrix} a & b & c \\ x & y & z \\ l & m & n \end{vmatrix},$$

$$\begin{vmatrix} x^2 & y^2 & z^2 \\ x & y & z \\ 1 & 1 & 1 \end{vmatrix}, \quad \begin{vmatrix} x^2 & x & 1 \\ y^2 & y & 1 \\ z^2 & z & 1 \end{vmatrix}, \quad \begin{vmatrix} x^2 & a & l \\ y^2 & b & m \\ z^2 & c & n \end{vmatrix}.$$

2. Expand the six determinants of Example 1 by their first columns.

3. Prove that the following equations are true by expanding each determinant separately:

$$\begin{vmatrix} ax & x & a \\ by & y & b \\ cz & z & c \end{vmatrix} = \begin{vmatrix} yz & zx & xy \\ 1 & 1 & 1 \\ bc & ca & ab \end{vmatrix},$$

$$\begin{vmatrix} ax & by & cz \\ a^2 & b^2 & c^2 \\ 1 & 1 & 1 \end{vmatrix} = \begin{vmatrix} x & y & z \\ a & b & c \\ bc & ca & ab \end{vmatrix},$$

$$\begin{vmatrix} x & y & z \\ a & b & c \\ a^2 & b^2 & c^2 \end{vmatrix} = \begin{vmatrix} bcx & cay & abz \\ 1 & 1 & 1 \\ a & b & c \end{vmatrix}.$$

5. Properties of determinants of order three†

The main properties are those given in § 3 for determinants of order two, though in some instances the statement for the latter has to be generalized a little. We number the properties in the same way as before and we take as our standard determinant for the purposes of our proofs

$$\begin{vmatrix} a_1 & b_1 & c_1 \\ a_2 & b_2 & c_2 \\ a_3 & b_3 & c_3 \end{vmatrix}.$$

When we wish to refer to a determinant without writing it in full we shall denote it by Δ.

I. The determinant has three rows and three columns; the individual letters $a_1, b_1, \ldots, c_3$ are called the ELEMENTS of Δ.

II. *The value of the determinant is unaltered if rows become columns and columns become rows.* That is,

$$\begin{vmatrix} a_1 & b_1 & c_1 \\ a_2 & b_2 & c_2 \\ a_3 & b_3 & c_3 \end{vmatrix} = \begin{vmatrix} a_1 & a_2 & a_3 \\ b_1 & b_2 & b_3 \\ c_1 & c_2 & c_3 \end{vmatrix},$$

an equality that is best proved (at this stage) by expanding the one determinant by its first row, the other by its first column, and seeing that the two expansions are the same.

† Some readers may prefer to omit § 5, the details of which appeal to the intending mathematician rather than to the scientist. The latter may profitably 'read it over', for the results are important.

III. *The interchange of any two columns, or of any two rows, multiplies the value of the determinant by* -1.

We may prove that

$$\begin{vmatrix} a_1 & b_1 & c_1 \\ a_2 & b_2 & c_2 \\ a_3 & b_3 & c_3 \end{vmatrix} = -\begin{vmatrix} c_1 & b_1 & a_1 \\ c_2 & b_2 & a_2 \\ c_3 & b_3 & a_3 \end{vmatrix},$$

in which the first and third columns are interchanged, by noting that, on expanding the determinants by their first rows, the L.H.S. is

$$a_1(b_2 c_3 - b_3 c_2) - b_1(a_2 c_3 - a_3 c_2) + c_1(a_2 b_3 - a_3 b_2) \tag{1}$$

and the R.H.S. is

$$-c_1(b_2 a_3 - b_3 a_2) + b_1(c_2 a_3 - c_3 a_2) - a_1(c_2 b_3 - c_3 b_2),$$

which is equal to

$$c_1(a_2 b_3 - a_3 b_2) - b_1(a_2 c_3 - a_3 c_2) + a_1(b_2 c_3 - b_3 c_2), \tag{2}$$

and (2) contains the same terms as (1), but in a different order.

The proofs for other interchanges are similar.

IV. *If a determinant has two columns, or two rows, identical, its value is zero.*

PROOF. Suppose that the value of the determinant is V. Then

(i) if the two identical columns (or rows) are interchanged, the determinant is unaltered and its value remains V;

(ii) if the two identical columns (or rows) are interchanged, the value V is multiplied by -1 and becomes $-V$ (by property III).

Hence $V = -V$ and so $2V = 0$.

V. *If each element of one column, or row, is multiplied by a factor K, the value of the determinant is thereby multiplied by K. For example,*

$$\begin{vmatrix} 2a & l & x \\ 2b & m & y \\ 2c & n & z \end{vmatrix} = 2\begin{vmatrix} a & l & x \\ b & m & y \\ c & n & z \end{vmatrix}.$$

This property is proved by expanding the determinants; in the example given, expand by the first column.

COROLLARY TO IV AND V. *If the elements of one column (or row) are K times the elements of another column (or row), the value of the determinant is zero. For example,*

$$\begin{vmatrix} x & a & 3a \\ y & b & 3b \\ z & c & 3c \end{vmatrix} = 0, \qquad \begin{vmatrix} u & v & w \\ 4l & 4m & 4n \\ l & m & n \end{vmatrix} = 0.$$

VI. *The value of a determinant is unaltered if to each element of one column (or row) is added the same multiple of the corresponding element of another column (or row); for example,*

$$\begin{vmatrix} a_1 & b_1 & c_1 \\ a_2 & b_2 & c_2 \\ a_3 & b_3 & c_3 \end{vmatrix} = \begin{vmatrix} a_1+lb_1 & b_1 & c_1 \\ a_2+lb_2 & b_2 & c_2 \\ a_3+lb_3 & b_3 & c_3 \end{vmatrix}.$$

PROOF. We shall prove the example given in the enunciation; other additions may be treated in a similar way. [Columns must not be added to rows.]

The determinant on the right is (by III) equal to

$$-\begin{vmatrix} c_1 & b_1 & a_1+lb_1 \\ c_2 & b_2 & a_2+lb_2 \\ c_3 & b_3 & a_3+lb_3 \end{vmatrix}$$

$$= -c_1\begin{vmatrix} b_2 & a_2+lb_2 \\ b_3 & a_3+lb_3 \end{vmatrix} + c_2\begin{vmatrix} b_1 & a_1+lb_1 \\ b_3 & a_3+lb_3 \end{vmatrix} - c_3\begin{vmatrix} b_1 & a_1+lb_1 \\ b_2 & a_2+lb_2 \end{vmatrix}.$$

By property VI of determinants of order two (§ 3), this equals

$$-c_1\begin{vmatrix} b_2 & a_2 \\ b_3 & a_3 \end{vmatrix} + c_2\begin{vmatrix} b_1 & a_1 \\ b_3 & a_3 \end{vmatrix} - c_3\begin{vmatrix} b_1 & a_1 \\ b_2 & a_2 \end{vmatrix}$$

$$= -\begin{vmatrix} c_1 & b_1 & a_1 \\ c_2 & b_2 & a_2 \\ c_3 & b_3 & a_3 \end{vmatrix} = \begin{vmatrix} a_1 & b_1 & c_1 \\ a_2 & b_2 & c_2 \\ a_3 & b_3 & c_3 \end{vmatrix},$$

which proves our result.

NOTE. The essential point of the foregoing proof is the way in which property VI for two rows and columns is used to prove property VI for three rows and columns. The other details are relatively unimportant; they should not be memorized.

COROLLARIES TO VI. In the following l, m, n denote any numbers, positive or negative: the results given are often useful in evaluating determinants.

(i) $$\begin{vmatrix} a_1 & b_1 & c_1 \\ a_2 & b_2 & c_2 \\ a_3 & b_3 & c_3 \end{vmatrix} = \begin{vmatrix} a_1+lb_1+mc_1 & b_1+nc_1 & c_1 \\ a_2+lb_2+mc_2 & b_2+nc_2 & c_2 \\ a_3+lb_3+mc_3 & b_3+nc_3 & c_3 \end{vmatrix},$$

that is, *we can add multiples of columns that 'come after'*;

(ii) $$\begin{vmatrix} a_1 & b_1 & c_1 \\ a_2 & b_2 & c_2 \\ a_3 & b_3 & c_3 \end{vmatrix} = \begin{vmatrix} a_1 & b_1+la_1 & c_1+mb_1+na_1 \\ a_2 & b_2+la_2 & c_2+mb_2+na_2 \\ a_3 & b_3+la_3 & c_3+mb_3+na_3 \end{vmatrix},$$

that is, *we can add multiples of columns that 'come before'*.

Similarly, we can add to the first row multiples of the second and third rows and add to the second row multiples of the third; or we can add to any row multiples of rows that 'come before'.

PROOF OF (i). The proof consists of successive applications of VI. Thus, by using VI at each step,

$$\begin{vmatrix} a_1 & b_1 & c_1 \\ a_2 & b_2 & c_2 \\ a_3 & b_3 & c_3 \end{vmatrix} = \begin{vmatrix} a_1+lb_1 & b_1 & c_1 \\ a_2+lb_2 & b_2 & c_2 \\ a_3+lb_3 & b_3 & c_3 \end{vmatrix}$$

$$= \begin{vmatrix} a_1+lb_1+mc_1 & b_1 & c_1 \\ a_2+lb_2+mc_2 & b_2 & c_2 \\ a_3+lb_3+mc_3 & b_3 & c_3 \end{vmatrix}$$

$$= \begin{vmatrix} a_1+lb_1+mc_1 & b_1+nc_1 & c_1 \\ a_2+lb_2+mc_2 & b_2+nc_2 & c_2 \\ a_3+lb_3+mc_3 & b_3+nc_3 & c_3 \end{vmatrix}.$$

NOTATION FOR VI AND ITS COROLLARIES. *Worked examples.*

We use the same notation as we did in § 3. Examples of the use of the notation are:

EXAMPLE 1. *Find the value of*

$$\Delta = \begin{vmatrix} 87 & 42 & 3 \\ 45 & 18 & 7 \\ 59 & 28 & 3 \end{vmatrix}.$$

SOLUTION.

$$\Delta = \begin{vmatrix} 0 & 36 & 3 \\ 2 & 4 & 7 \\ 0 & 22 & 3 \end{vmatrix} \qquad \begin{matrix} c_1' = c_1-2c_2-c_3, \\ c_2' = c_2-2c_3, \end{matrix}$$

[the notation on the right indicates how the new first and second columns are formed]

$$= -2\begin{vmatrix} 36 & 3 \\ 22 & 3 \end{vmatrix}$$

$$= -2(108-66) = -84.$$

NOTE. The numbers first given are too high for ease of calculation, so we subtract convenient multiples of 'columns that come after'. If we can choose multiples so as to get one or more 0's in the first column, the subsequent calculation is made easier.

The first column of the new determinant is

$$0 = 87-2.42-3,$$
$$2 = 45-2.18-7,$$
$$0 = 59-2.28-3.$$

EXAMPLE 2. *Prove that*

$$\Delta = \begin{vmatrix} 0 & a-b & a-c \\ b-a & 0 & b-c \\ c-a & c-b & 0 \end{vmatrix} = 0.$$

SOLUTION. Form a new determinant by taking

$$r_1' = r_1-r_2, \qquad r_2' = r_2-r_3,$$

i.e. new first row = first row of Δ—second row of Δ,
new second row = second row of Δ—third row of Δ.

The new first row is thus

$$0-(b-a) = a-b, \quad (a-b)-0 = a-b, \quad (a-c)-(b-c) = a-b,$$

and the given determinant is equal to

$$\begin{vmatrix} a-b & a-b & a-b \\ b-c & b-c & b-c \\ c-a & c-b & 0 \end{vmatrix}.$$

Remove the factor $a-b$ from the first row and the factor $b-c$ from the second. This gives (by V)

$$\Delta = (a-b)(b-c)\begin{vmatrix} 1 & 1 & 1 \\ 1 & 1 & 1 \\ c-a & c-b & 0 \end{vmatrix},$$

which is zero because two rows are identical (IV).

EXAMPLE 3. *Prove that*

$$\Delta = \begin{vmatrix} 1 & 1 & 1 \\ 1 & r & r^2 \\ 1 & r^2 & r^4 \end{vmatrix} = r(r-1)^2(r^2-1).$$

SOLUTION.

On taking $c_2' = c_2-c_1$, $c_3' = c_3-c_1$,

$$\Delta = \begin{vmatrix} 1 & 0 & 0 \\ 1 & r-1 & r^2-1 \\ 1 & r^2-1 & r^4-1 \end{vmatrix}.$$

When we expand this by its first row we get

$$\Delta = \begin{vmatrix} r-1 & r^2-1 \\ r^2-1 & r^4-1 \end{vmatrix} \tag{A}$$

$$= (r-1)(r^2-1)\begin{vmatrix} 1 & 1 \\ r+1 & r^2+1 \end{vmatrix},$$

on removing the factor $r-1$ from the first column and r^2-1 from the second column of (A). Thus

$$\Delta = (r-1)(r^2-1)(r^2-r) = r(r-1)^2(r^2-1).$$

EXAMPLE 4. *Prove that*

$$\Delta = \begin{vmatrix} 19 & 17 & 15 \\ 9 & 8 & 7 \\ 1 & 1 & 1 \end{vmatrix} = 0.$$

SOLUTION. We notice that

$$19-2.9-1 = 0;$$
$$17-2.8-1 = 0,$$
$$15-2.7-1 = 0.$$

On taking $r_1' = r_1 - 2r_2 - r_3$ we obtain (by VI, Corollary)

$$\Delta = \begin{vmatrix} 0 & 0 & 0 \\ 9 & 8 & 7 \\ 1 & 1 & 1 \end{vmatrix},$$

which is equal to zero.

[The expansion of this determinant by its first row is

$$0 \times \begin{vmatrix} 8 & 7 \\ 1 & 1 \end{vmatrix} - 0 \times \begin{vmatrix} 9 & 7 \\ 1 & 1 \end{vmatrix} + 0 \times \begin{vmatrix} 9 & 8 \\ 1 & 1 \end{vmatrix},$$

and so its value is zero. In fact any determinant with a complete row or column of zeros is itself zero.]

Examples XIII d

1. Find the values of the determinants

$$\begin{vmatrix} 96 & 47 & 23 \\ 97 & 43 & 21 \\ 98 & 49 & 25 \end{vmatrix}, \quad \begin{vmatrix} 15 & 7 & 8 \\ 17 & 8 & 9 \\ 28 & 9 & 12 \end{vmatrix}, \quad \begin{vmatrix} 8 & 9 & 10 \\ 4 & 5 & 6 \\ 91 & 92 & 93 \end{vmatrix}.$$

2. Find the values of the determinants

$$\begin{vmatrix} 17 & 46 & 7 \\ 20 & 49 & 8 \\ 23 & 52 & 9 \end{vmatrix}, \quad \begin{vmatrix} 17 & 46 & 7 \\ 20 & 49 & 8 \\ 23 & 52 & 12 \end{vmatrix}, \quad \begin{vmatrix} 1001 & 17 & 2 \\ 1002 & 18 & 4 \\ 1003 & 19 & 7 \end{vmatrix}.$$

3. Prove, by adding or subtracting multiples of rows, that each of the determinants

$$\begin{vmatrix} y+z & z+x & x+y \\ x & y & z \\ 1 & 1 & 1 \end{vmatrix}, \quad \begin{vmatrix} b-c & c-a & a-b \\ c & a & b \\ b & c & a \end{vmatrix}, \quad \begin{vmatrix} x+y & u+v & X+Y \\ 2x+3y & 2u+3v & 2X+3Y \\ 4x+5y & 4u+5v & 4X+5Y \end{vmatrix},$$

is equal to zero.

4. Prove, by adding or subtracting multiples of columns, that each of the determinants

$$\begin{vmatrix} y-z & y+z & y \\ z-x & z+x & z \\ x-y & x+y & x \end{vmatrix}, \quad \begin{vmatrix} (x-1)^2 & x^2+1 & x \\ (y-1)^2 & y^2+1 & y \\ (z-1)^2 & z^2+1 & z \end{vmatrix}, \quad \begin{vmatrix} x+y & 7x+8y & 9x+11y \\ u+v & 7u+8v & 9u+11v \\ X+Y & 7X+8Y & 9X+11Y \end{vmatrix}$$

is equal to zero.

5. Prove that

(i) $$\begin{vmatrix} 1 & 1 & 1 \\ x & y & z \\ x^2 & y^2 & z^2 \end{vmatrix} = (y-z)(z-x)(x-y),$$

(ii) $$\begin{vmatrix} 1 & r & r^2 \\ 1 & r^2 & r^4 \\ 1 & r^3 & r^6 \end{vmatrix} = r^4(r-1)^2(r^2-1),$$

(iii) $$\begin{vmatrix} 1 & 1 & 1 \\ 1 & r^2 & r^4 \\ 1 & r^3 & r^6 \end{vmatrix} = r^2(r-1)(r^2-1)(r^3-1).$$

6. Prove that

$$\begin{vmatrix} x+y & 7x+8y & 9x+10y \\ u+v & 7u+8v & 9u+11v \\ X+Y & 7X+8Y & 9X+12Y \end{vmatrix} = v(xY+yX)-2uyY.$$

HINT. Simplify the determinant by subtracting multiples of columns before expanding.

7. Prove, by using VI, Corollary, that *a determinant whose third row is the sum of the first and second rows has the value zero.*

8. Prove that $$\begin{vmatrix} x & lx+mu & u \\ y & ly+mv & v \\ z & lz+mw & w \end{vmatrix} = 0.$$

6.* Minors and cofactors

6.1. The determinant of order two obtained by deleting one row and one column from a determinant of order three is called a MINOR. For example,

$$\begin{vmatrix} a_1 & c_1 \\ a_3 & c_3 \end{vmatrix} \tag{1}$$

is obtained by deleting the second row and the second column from

$$\Delta \equiv \begin{vmatrix} a_1 & b_1 & c_1 \\ a_2 & b_2 & c_2 \\ a_3 & b_3 & c_3 \end{vmatrix},$$

and, since the deleted row and column both contain the element b_2, we call (1) the minor of b_2 in Δ.

6.2. The expansion of Δ is

$$a_1 b_2 c_3 - a_1 b_3 c_2 + a_2 b_3 c_1 - a_2 b_1 c_3 + a_3 b_1 c_2 - a_3 b_2 c_1. \tag{2}$$

The terms of (2) may be grouped together in many different ways; two such ways are

$$a_1 \times (\text{minor of } a_1) - a_2 \times (\text{minor of } a_2) + a_3 \times (\text{minor of } a_3), \tag{3}$$

$$-a_2 \times (\text{minor of } a_2) + b_2 \times (\text{minor of } b_2) - c_2 \times (\text{minor of } c_2). \tag{4}$$

The presence of the plus and minus signs makes these expressions awkward to handle. We can simplify these expressions by introducing the 'SIGNED MINORS' or the 'COFACTORS', which are merely the minors prefixed by the appropriate signs.

DEFINITION. *The cofactor of the element in the rth row and sth column of a determinant Δ is $(-1)^{r+s}$ times the minor of that element.*

For example,

a_1 is in the first row ($r = 1$) and first column ($s = 1$) and the cofactor of a_1 is

$$(-1)^{1+1}\begin{vmatrix} b_2 & c_2 \\ b_3 & c_3 \end{vmatrix} = \begin{vmatrix} b_2 & c_2 \\ b_3 & c_3 \end{vmatrix};$$

b_3 is in the third row ($r = 3$) and second column ($s = 2$) and the cofactor of b_3 is

$$(-1)^{3+2}\begin{vmatrix} a_1 & c_1 \\ a_2 & c_2 \end{vmatrix} = -\begin{vmatrix} a_1 & c_1 \\ a_2 & c_2 \end{vmatrix}.$$

NOTATION. We denote the cofactor of a_1 by A_1, the cofactor of b_3 by B_3, and so on. We use a similar notation with other determinants.

6.3. *The expansion of a determinant.* When we use the above notation for the cofactors, the various ways of expanding a determinant can be set out in a very simple form.

We know that

$$\Delta = \begin{vmatrix} a_1 & b_1 & c_1 \\ a_2 & b_2 & c_2 \\ a_3 & b_3 & c_3 \end{vmatrix}$$

$$= a_1 b_2 c_3 - a_1 b_3 c_2 + a_2 b_3 c_1 - a_2 b_1 c_3 + a_3 b_1 c_2 - a_3 b_2 c_1.$$

This may be written in any one of the forms

$$\left.\begin{aligned} \Delta &= a_1 A_1 + b_1 B_1 + c_1 C_1, & \Delta &= a_1 A_1 + a_2 A_2 + a_3 A_3, \\ \Delta &= a_2 A_2 + b_2 B_2 + c_2 C_2, & \Delta &= b_1 B_1 + b_2 B_2 + b_3 B_3, \\ \Delta &= a_3 A_3 + b_3 B_3 + c_3 C_3, & \Delta &= c_1 C_1 + c_2 C_2 + c_3 C_3. \end{aligned}\right\} \quad (5)$$

The result may be expressed as

Take any row of a determinant, multiply each element of the row by its own cofactor and add; the sum is the value of the determinant; it is called the expansion of the determinant by that row.

Take any column of a determinant, multiply each element of the column by its own cofactor and add; the sum is the value of the determinant; it is called the expansion of the determinant by that column.

6.4. *Further properties of cofactors.* Consider the determinant

$$\Delta' = \begin{vmatrix} a_2 & b_2 & c_2 \\ a_2 & b_2 & c_2 \\ a_3 & b_3 & c_3 \end{vmatrix}.$$

We may note two things,

(i) $\Delta' = 0$ because two rows are identical;

(ii) Δ' differs from Δ only in having a_2, b_2, c_2 instead of a_1, b_1, c_1 in the first row. Hence, when we expand Δ' by its first row we get (since $\Delta = a_1 A_1 + b_1 B_1 + c_1 C_1$)

$$\Delta' = a_2 A_1 + b_2 B_1 + c_2 C_1.$$

This shows that $$0 = a_2 A_1 + b_2 B_1 + c_2 C_1, \tag{6}$$

and, in the same way, we may show that

'*the sum of elements of any one row multiplied by the corresponding cofactors of* ANOTHER *row is zero*'.

As further examples we note that

$$\left.\begin{aligned} a_1 A_3 + b_1 B_3 + c_1 C_3 &= 0, \\ a_2 A_3 + b_2 B_3 + c_2 C_3 &= 0, \\ a_3 A_1 + b_3 B_1 + c_3 C_1 &= 0. \end{aligned}\right\} \tag{7}$$

Similarly, we may show that

'*the sum of elements of a column multiplied by the corresponding cofactors of* ANOTHER *column is zero*'. For example,

$$\left.\begin{aligned} a_1 B_1 + a_2 B_2 + a_3 B_3 &= 0, \\ b_1 A_1 + b_2 A_2 + b_3 A_3 &= 0, \\ c_1 A_1 + c_2 A_2 + c_3 A_3 &= 0. \end{aligned}\right\} \tag{8}$$

6.5. The propositions contained in §§ 6.3 and 6.4 are of the first importance in nearly all work involving determinants and the reader should know them thoroughly. He should also know how to apply the notation for cofactors and the foregoing results to any determinant he may encounter. The determinant

$$\Delta = \begin{vmatrix} a & h & g \\ h & b & f \\ g & f & c \end{vmatrix}$$

is of particular importance in analytical geometry. Here A, H,

$G,\ldots$ denote the cofactors of $a, h, g,\ldots$ in Δ. For example,†

$$B = (-1)^{2+2}\begin{vmatrix} a & g \\ g & c \end{vmatrix} = ac-g^2;$$

$$H = (-1)^{2+1}\begin{vmatrix} h & g \\ f & c \end{vmatrix} = fg-ch;$$

the complete list of cofactors being

$$A = bc-f^2, \qquad F = gh-af,$$
$$B = ca-g^2, \qquad G = hf-bg,$$
$$C = ab-h^2, \qquad H = fg-ch.$$

The expansion of Δ is

$$abc+2fgh-af^2-bg^2-ch^2; \tag{9}$$

the formulae of § 6.3 give

$$\Delta = aA+hH+gG,$$
$$\Delta = hH+bB+fF,$$
$$\Delta = gG+fF+cC.$$

The reader will notice the symmetry of rows and columns in Δ; because of this symmetry the expansions of Δ by rows are exactly the same in form as the expansions of Δ by columns.

Further, the propositions of § 6.4 that 'the sum of elements of one row (or column) multiplied by the corresponding cofactors of another row (or column) is zero' give results like

$$aG+hF+gC = 0, \qquad hA+bH+fG = 0. \tag{10}$$

If the reader can, in his analytical geometry of 'the general conic', learn to use expansions like $aA+hH+gG$ instead of (9), and to recognize results like (10) instead of laboriously cancelling terms in

$$a(hf-bg)+h(gh-af)+g(ab-h^2)$$

and $$h(bc-f^2)+b(fg-ch)+f(hf-bg),$$

he will have made a great stride in the *use* of determinants.

7. The solution of linear equations

7.1. There are two major results. These are:

† Notice that the minor of h is the same whether we consider the h of the first row or of the first column.

I. *The solution of the three simultaneous equations*

$$a_1 x+b_1 y+c_1 z = d_1,$$
$$a_2 x+b_2 y+c_2 z = d_2,$$
$$a_3 x+b_3 y+c_3 z = d_3,$$

is given, in general, by

$$\Delta x = \begin{vmatrix} d_1 & b_1 & c_1 \\ d_2 & b_2 & c_2 \\ d_3 & b_3 & c_3 \end{vmatrix}, \qquad \Delta y = \begin{vmatrix} a_1 & d_1 & c_1 \\ a_2 & d_2 & c_2 \\ a_3 & d_3 & c_3 \end{vmatrix}, \qquad \Delta z = \begin{vmatrix} a_1 & b_1 & d_1 \\ a_2 & b_2 & d_2 \\ a_3 & b_3 & d_3 \end{vmatrix},$$

where Δ *denotes the determinant whose elements are the coefficients of* x, y, z, *namely*

$$\Delta = \begin{vmatrix} a_1 & b_1 & c_1 \\ a_2 & b_2 & c_2 \\ a_3 & b_3 & c_3 \end{vmatrix}.$$

The exception to the general rule occurs when $\Delta = 0$.

It will be noticed that a column of d's replaces a column of a's in the first, a column of b's in the second, and a column of c's in the third determinant of the solution.

II. *If the equations*

$$a_1 x+b_1 y+c_1 z = 0,$$
$$a_2 x+b_2 y+c_2 z = 0,$$
$$a_3 x+b_3 y+c_3 z = 0,$$

are satisfied by any set of values of x, y, z *other than* $x = y = z = 0$, *then* $\Delta = 0$.

Conversely, if $\Delta = 0$, *then the three equations are satisfied by a set of values of* x, y, z *other than* $x = y = z = 0$.

It will be noticed that $x = y = z = 0$ always satisfies these equations. The point at issue is whether there is another solution.

COROLLARY. *If* $\Delta \neq 0$, *the equations have no solution other than* $x = 0$, $y = 0$, $z = 0$.

In such a case the first two equations are satisfied by $x = \lambda(b_1 c_2 - b_2 c_1)$, $y = -\lambda(a_1 c_2 - a_2 c_1)$, $z = \lambda(a_1 b_2 - a_2 b_1)$ for any value of λ, but these values will satisfy the third equation only if $\lambda = 0$.

All readers should take careful note of these results, which are widely used. Many readers will find the proofs, which are given in the subsections that follow, not altogether easy. The proofs need not be mastered thoroughly on a first reading of the book; when the reader has grown familiar with determinants and cofactors, he will find that the proofs fall naturally into place; he will then wonder where his difficulty lay and probably consider the proofs presented here to be unnecessarily long and explanatory.

7.2.* *Proof of I.* Suppose that $\Delta \neq 0$. Then, using the notation of cofactors of Δ, the solution given may be written

$$\Delta x = d_1 A_1 + d_2 A_2 + d_3 A_3, \qquad \Delta y = d_1 B_1 + d_2 B_2 + d_3 B_3,$$
$$\Delta z = d_1 C_1 + d_2 C_2 + d_3 C_3.$$

If we substitute these values in $a_1 x + b_1 y + c_1 z$, we get

$$\{a_1(d_1 A_1 + d_2 A_2 + d_3 A_3) + b_1(d_1 B_1 + d_2 B_2 + d_3 B_3) +$$
$$+ c_1(d_1 C_1 + d_2 C_2 + d_3 C_3)\}/\Delta$$
$$= \frac{d_1}{\Delta}(a_1 A_1 + b_1 B_1 + c_1 C_1) + \frac{d_2}{\Delta}(a_1 A_2 + b_1 B_2 + c_1 C_2) +$$
$$+ \frac{d_3}{\Delta}(a_1 A_3 + b_1 B_3 + c_1 C_3).$$

Now, by the results of §§ 6.3. and 6.4,

$$a_1 A_1 + b_1 B_1 + c_1 C_1 = \Delta, \qquad a_1 A_2 + b_1 B_2 + c_1 C_2 = 0,$$
$$a_1 A_3 + b_1 B_3 + c_1 C_3 = 0,$$

and so, with the above values of x, y, z,

$$a_1 x + b_1 y + c_1 z = d_1,$$

and we may show in a similar way that these values satisfy the other two equations.

Note on the proof of I. In order to *find* the solution, for x say, multiply the given equations by A_1, A_2, A_3 respectively and add: the result is, in virtue of §§ 6.3 and 6.4,

$$\Delta . x + 0 . y + 0 . z = d_1 A_1 + d_2 A_2 + d_3 A_3.$$

Similarly,

$$0 . x + \Delta . y + 0 . z = d_1 B_1 + d_2 B_2 + d_3 B_3,$$
$$0 . x + 0 . y + \Delta . z = d_1 C_1 + d_2 C_2 + d_3 C_3.$$

7.3. When $\Delta = 0$ the solution breaks down because we cannot then divide by Δ. In such a case it will sometimes be found that the three equations are not independent and that one of them is automatically satisfied whenever the other two are, sometimes that the equations have no solution. We shall not attempt to handle the general theory, but we show the sort of thing that happens by considering two particular examples.

EXAMPLE 1: $\Delta = 0$. *Solve the equations*

$$2x+y-z = 7, \tag{i}$$

$$5x-4y+7z = 1, \tag{ii}$$

$$7x-3y+6z = 8. \tag{iii}$$

SOLUTION. In this example

$$\Delta = \begin{vmatrix} 2 & 1 & -1 \\ 5 & -4 & 7 \\ 7 & -3 & 6 \end{vmatrix} = 0$$

[the third row is the sum of the other two rows].

If values of x, y, z satisfy (i) and (ii), then also

$$(2x+y-z-7)+(5x-4y+7z-1) = 0,$$

i.e.
$$7x-3y+6z = 8,$$

and (iii) is automatically satisfied.

Hence our problem reduces to that of solving (i) and (ii). Let z have any value whatsoever, say $z = \lambda$. Then we have to solve, for x and y, the two equations

$$2x+y = 7+\lambda,$$

$$5x-4y = 1-7\lambda.$$

The solution is $13x = 29-3\lambda$, $13y = 33+19\lambda$. Thus the solution of the three given equations is

$$13x = 29-3\lambda, \qquad 13y = 33+19\lambda, \qquad z = \lambda,$$

where λ is arbitrary.

EXAMPLE 2: $\Delta = 0$. *Solve the equations*

$$x+y+z = 3, \tag{i}$$

$$2x-3y+2z = 1, \tag{ii}$$

$$3x-2y+3z = 7. \tag{iii}$$

SOLUTION. In this example

$$\Delta = \begin{vmatrix} 1 & 1 & 1 \\ 2 & -3 & 2 \\ 3 & -2 & 3 \end{vmatrix} = 0$$

[two columns are identical].

If values of x, y, z are chosen to satisfy (i) and (ii), then also

$$(x+y+z-3)+(2x-3y+2z-1) = 0,$$

i.e.
$$3x-2y+3z = 4,$$

and so (iii) cannot be satisfied by any values of x, y, z that satisfy (i) and (ii).

The three equations are *inconsistent* and no set of values of x, y, z can be found to satisfy all of them simultaneously.

7.4.* *Proof of* II (i). We here prove that '*if the equations*

$$a_1 x+b_1 y+c_1 z = 0,$$
$$a_2 x+b_2 y+c_2 z = 0,$$
$$a_3 x+b_3 y+c_3 z = 0,$$

are satisfied by a set of values of x, y, z other than $x = y = z = 0$, then $\Delta = 0$'.

As before, let A_1, B_2,... denote the cofactors of a_1, b_2,... in

$$\Delta = \begin{vmatrix} a_1 & b_1 & c_1 \\ a_2 & b_2 & c_2 \\ a_3 & b_3 & c_3 \end{vmatrix}.$$

Multiply the given equations by A_1, A_2, A_3 respectively and add. The result is

$$\Delta . x = 0$$

(as in § 7.2). Similarly, we obtain $\Delta . y = 0$ and $\Delta . z = 0$.

Hence, if any set of values of x, y, z satisfies the three given equations, this set of values also satisfies

$$\Delta . x = 0, \qquad \Delta . y = 0, \qquad \Delta . z = 0,$$

and if one or more of x, y, z differs from zero, it follows that $\Delta = 0$.

7.5.* *Proof of* II (ii). We here prove that '*if* $\Delta = 0$, *then the equations*

$$a_1 x + b_1 y + c_1 z = 0,$$
$$a_2 x + b_2 y + c_2 z = 0,$$
$$a_3 x + b_3 y + c_3 z = 0,$$

are satisfied by a set of values of x, y, z *other than the set*

$$x = y = z = 0\text{'}.$$

The run of the proof depends on whether

(*a*) one at least of the cofactors A_1, B_1,..., C_3 of Δ is not equal to zero,

or (*b*) all the cofactors of Δ are zero.

We are given that $\Delta = 0$.

FIRST suppose that one at least of the cofactors differs from zero; say, $B_2 \neq 0$. Then, for an arbitrary value of k,

$$x = kA_2, \qquad y = kB_2, \qquad z = kC_2 \tag{1}$$

satisfies all three equations; for

$$a_1 A_2 + b_1 B_2 + c_1 C_2 = 0 \qquad \text{(by § 6.4)},$$
$$a_2 A_2 + b_2 B_2 + c_2 C_2 = \Delta = 0 \quad \text{(by hypothesis)},$$
$$a_3 A_2 + b_3 B_2 + c_3 C_2 = 0 \qquad \text{(by § 6.4)}.$$

Moreover, when $k \neq 0$ the value $y = kB_2$ differs from zero. This proves the proposition when $B_2 \neq 0$ and it can be similarly proved when any one of the cofactors is not equal to zero.

NEXT suppose that all the cofactors are zero, but that one at least of the coefficients a_1,..., c_3 is not zero; say, $c_1 \neq 0$.

Let x, y have arbitrary non-zero values; say,

$$x = c_1 k, \qquad y = c_1 l \quad (k, l \neq 0),$$

and choose z so that the first equation is satisfied. That is,

$$x = c_1 k, \qquad y = c_1 l, \qquad z = -(a_1 k + b_1 l). \tag{2}$$

When we substitute these values in $a_2 x + b_2 y + c_2 z$ we obtain

$$a_2 c_1 k + b_2 c_1 l - c_2 a_1 k - c_2 b_1 l = k(a_2 c_1 - a_1 c_2) + l(b_2 c_1 - b_1 c_2)$$
$$= kB_3 - lA_3 = 0,$$

by the hypothesis that all the cofactors are zero.

Similarly,

$$a_3 c_1 k + b_3 c_1 l - c_3(a_1 k + b_1 l) = 0.$$

Hence the values (2) satisfy all three equations; moreover, $c_1 k \neq 0$ by hypothesis. This proves the proposition when $c_1 \neq 0$ and it can be similarly proved when any one of the coefficients $a_1, \ldots, c_3$ is not zero.

FINALLY suppose that all the coefficients are zero. Then the three 'equations' are satisfied by any values of x, y, z whatsoever and the proposition is proved in this (very easy) case also.

Hence, in all circumstances, 'if $\Delta = 0$, the three equations are satisfied by a set of values of x, y, z which are not all zero'.

Notice that the proofs show how to write down values of x, y, z that satisfy the equations.

7.6.* *A deduction from* II. The equations

$$\left.\begin{aligned} a_1 x + b_1 y + c_1 &= 0, \\ a_2 x + b_2 y + c_2 &= 0, \\ a_3 x + b_3 y + c_3 &= 0, \end{aligned}\right\} \tag{3}$$

are obtained from the three equations of § 7.5 by putting $z = 1$.

Thus, if the equations (3) are *consistent*, that is, if there are values of x and y that satisfy them simultaneously, the equations of § 7.5 have a solution with $z = 1$; and, therefore, $\Delta = 0$.

If $\Delta \neq 0$, the equations are *inconsistent*, that is, no pair of values of x, y satisfies all three equations simultaneously.

7.7.* *A theorem in analytical geometry.* The most important application of the ideas in § 7.6 occurs in analytical geometry. A theorem of frequent application in analytical geometry runs as follows:

THEOREM. *Let*

$$a_1 x + b_1 y + c_1 = 0, \tag{i}$$

$$a_2 x + b_2 y + c_2 = 0, \tag{ii}$$

$$a_3 x + b_3 y + c_3 = 0 \tag{iii}$$

be the equations of three different lines. Then, if the lines are concurrent or are all parallel, $\Delta = 0$; conversely, if $\Delta = 0$, the lines are either concurrent or are all parallel.

PROOF. If the lines are concurrent, there is a pair of values x, y that satisfies all three equations and, as in § 7.6, $\Delta = 0$.

If the lines are all parallel, $C_1 = a_2 b_3 - a_3 b_2 = 0$, and, similarly, $C_2 = C_3 = 0$, so that

$$\Delta = c_1 C_1 + c_2 C_2 + c_3 C_3 = 0.$$

Conversely, let $\Delta = 0$. Then, by II of § 7.1, there are constants l, m, n, not all zero, such that

$$la_1 + ma_2 + na_3 = 0,$$
$$lb_1 + mb_2 + nb_3 = 0,$$
$$lc_1 + mc_2 + nc_3 = 0.$$

Suppose that n is not zero. Then, for all x and y,

$$n(a_3 x + b_3 y + c_3) \equiv -l(a_1 x + b_1 y + c_1) - m(a_2 x + b_2 y + c_2).$$

Now if m were zero, this would give

EITHER

$$n(a_3 x + b_3 y + c_3) \equiv -l(a_1 x + b_1 y + c_1) \quad (l \neq 0),$$

which would mean that (i) and (iii) were the same line, and this is ruled out by the hypothesis that the three lines are different,

OR $$n(a_3 x + b_3 y + c_3) \equiv 0 \quad (l = 0),$$

which would mean that (iii) was satisfied by every x and y, which again is ruled out by the hypothesis that (iii) is the equation of a line.

Accordingly, $m \neq 0$ and, similarly, $l \neq 0$. Thus there are constants l, m, n, *no one of which is zero*, such that, for all x and y,

$$l(a_1 x + b_1 y + c_1) + m(a_2 x + b_2 y + c_2) + n(a_3 x + b_3 y + c_3) \equiv 0.$$

Hence, if two of the lines meet at a finite point, the third line passes through that point and, if two of the lines are parallel [say, $a_2 = ka_3$ and $b_2 = kb_3$, so that (ii) and (iii) are parallel], the third is parallel to both [$la_1 = -(n+mk)a_3$ and $lb_1 = -(n+mk)b_3$, so that (i) is parallel to (iii)].

EXAMPLES XIII E

1.* Find the values of X, Y, Z, the cofactors of x, y, z in the determinants

$$\text{(i)}\ \begin{vmatrix} x & y & z \\ u & v & w \\ l & m & n \end{vmatrix}, \qquad \text{(ii)}\ \begin{vmatrix} a & b & c \\ x & y & z \\ p & q & r \end{vmatrix}, \qquad \text{(iii)}\ \begin{vmatrix} 1 & 1 & 1 \\ 2 & 3 & 4 \\ x & y & z \end{vmatrix}.$$

2.* Prove that, when Δ denotes the determinant

$$\begin{vmatrix} a & h & g \\ h & b & f \\ g & f & c \end{vmatrix}$$

and A, H,... denote the cofactors of a, h,... in Δ,

$$aG^2+2hFG+bF^2+2gGC+2fFC+cC^2 = C\Delta.$$

HINT. Use $aG+hF+gC = 0$, etc.

3.* Obtain the solution of the equations

$$ax+hy+g = 0, \qquad hx+by+f = 0$$

in the form

$$\frac{x}{G} = \frac{y}{F} = \frac{1}{C},$$

where G, F, C have the meanings assigned in Example 2.

Solve the following sets of simultaneous equations:

4. $x+y+z = 1,$
$2x+4y-3z = 9,$
$5x-4y+z = 0.$

5. $2x-y+z = 7,$
$3x+y-5z = 13,$
$x+y+z = 5.$

6. $x+y+z = 6,$
$5x-y+2z = 9,$
$3x+6y-5z = 0.$

7. $3x-y+4z = 13,$
$5x+y-3z = 5,$
$x-y+z = 3.$

8.* $x+y+z = 1$, $ax+by+cz = d$, $a^2x+b^2y+c^2z = d^2$.

HINT. Use the answer to Examples XIII D, 5 (i).

Prove that each of the following sets of simultaneous equations is inconsistent:

9. $x+y+z = 1,$
$2x+4y-3z = 9,$
$3x+5y-2z = 11.$

10. $2x-y+z = 7,$
$3x+y-5z = 13,$
$2x-6y+14z = 21.$

11. $x-3y+z = 1,$
$2x-5y+2z = 3,$
$5x-9y+5z = 10.$

12.* $x+y+z = 6,$
$5x-y+2z = 9,$
$15x-9y+3z = 7.$

HINT for 12; $\Delta = 0$ and therefore, by II of § 7.1, constants l, m, n, which are not all zero, can be found to satisfy

$$l+5m+15n = 0, \qquad l-m-9n = 0, \qquad l+2m+3n = 0.$$

Find $l:m:n$ from any two of these.

Aliter. Solve two of the equations for x, y in terms of z and substitute in the third.

Examples 10 and 11 can be worked in the same way.

Find a general solution of each of the following sets of equations (if necessary, use the hint for Examples 10–12).

13. $$x+3y+z = 5,$$ $$2x+7y-z = 8,$$ $$5x+18y-4z = 19.$$

14. $$2x-y+z = 7,$$ $$3x+y-5z = 13,$$ $$5x-4z = 20.$$

15.* $$x+y+z = 6,$$ $$5x-y+2z = 9,$$ $$38x-16y+11z = 39.$$

CHAPTER XIV*

FURTHER PROPERTIES OF DETERMINANTS

1. The expansion of a determinant

As we have seen already (e.g. Chap. XIII, § 6),

$$\begin{vmatrix} l_1 & m_1 & n_1 \\ l_2 & m_2 & n_2 \\ l_3 & m_3 & n_3 \end{vmatrix} \tag{1}$$

$$= +l_1 m_2 n_3 - l_1 m_3 n_2 + l_2 m_3 n_1 - l_2 m_1 n_3 + l_3 m_1 n_2 - l_3 m_2 n_1. \tag{2}$$

Now there are 6 ways of arranging the numbers 1, 2, 3 in different orders and it is just these different orders that occur among the suffixes in (2). Moreover, if we think of the numbers 1, 2, 3 arranged round a circle, as in the figure, the sign of a term in (2) is

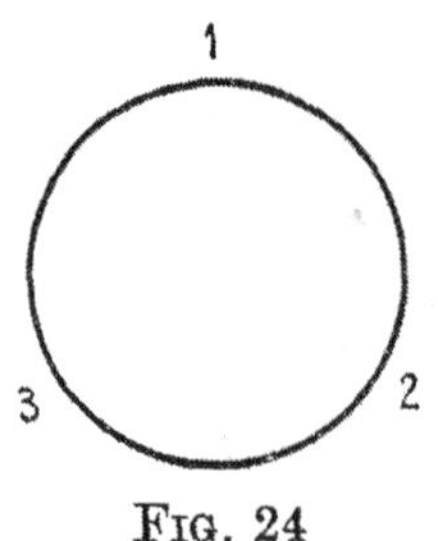

FIG. 24

(i) PLUS when the suffixes run clockwise,† 1 2 3, or 2 3 1, or 3 1 2;

(ii) MINUS when the suffixes run counter-clockwise, 1 3 2, or 2 1 3, or 3 2 1.

Accordingly, we may denote (2) by

$$\sum \pm \{lmn\}_{123}, \tag{3}$$

the notation implying the sum of the six terms got by putting the suffixes in their six possible orders and fixing the plus or minus sign according to the rule just given.

† This simple rule applies to determinants of order 3; neither it, nor any simple extension of it, applies to determinants of order 4 or more. The signs of the terms in such determinants are fixed by a rule that is not based on cyclic order.

2. Composite determinants

We apply the ideas of § 1 to the determinant

$$\begin{vmatrix} a_1+x_1 & b_1+y_1 & c_1+z_1 \\ a_2+x_2 & b_2+y_2 & c_2+z_2 \\ a_3+x_3 & b_3+y_3 & c_3+z_3 \end{vmatrix}, \tag{4}$$

in which each element is the sum of two terms.

Now the expansion of (4) is

$$\sum \pm\{(a+x)(b+y)(c+z)\}_{123}. \tag{5}$$

But the product
$$(a+x)(b+y)(c+z)$$
is the sum of the eight ($= 2^3$) terms

$$abc, \quad abz, \quad ayc, \quad ayz, \quad xbc, \quad xbz, \quad xyc, \quad xyz$$

and so (5) is the sum of the eight parts

$$\sum \pm\{abc\}_{123}, \qquad \sum \pm\{abz\}_{123}, \quad \ldots, \quad \sum \pm\{xyz\}_{123}. \tag{6}$$

Thus (4) is the sum of the eight determinants

$$\begin{vmatrix} a_1 & b_1 & c_1 \\ a_2 & b_2 & c_2 \\ a_3 & b_3 & c_3 \end{vmatrix}, \qquad \begin{vmatrix} a_1 & b_1 & z_1 \\ a_2 & b_2 & z_2 \\ a_3 & b_3 & z_3 \end{vmatrix}, \quad \ldots, \quad \begin{vmatrix} x_1 & y_1 & z_1 \\ x_2 & y_2 & z_2 \\ x_3 & y_3 & z_3 \end{vmatrix}.$$

We set out this result as a formal theorem.

Theorem 23. *The determinant*

$$\begin{vmatrix} a_1+x_1 & b_1+y_1 & c_1+z_1 \\ a_2+x_2 & b_2+y_2 & c_2+z_2 \\ a_3+x_3 & b_3+y_3 & c_3+z_3 \end{vmatrix}$$

is the sum of the 2^3 *determinants corresponding to the* 2^3 *different ways of choosing one letter from each column.*

Corollary 1. *If each element is the sum of three terms, the determinant can be expressed as the sum of* $3^3 = 27$ *determinants.*

There are 27 terms in the full-length expression of the product

$$(a+x+p)(b+y+q)(c+z+r).$$

Note. There is just one point in the sequel where we shall use this corollary. We shall NOT write down the 27 terms, but we shall have to know that they are there to be written.

COROLLARY 2. *The determinant*

$$\begin{vmatrix} a_1+x_1 & b_1+y_1 \\ a_2+x_2 & b_2+y_2 \end{vmatrix}$$

is the sum of the four ($= 2^2$) *determinants*

$$\begin{vmatrix} a_1 & b_1 \\ a_2 & b_2 \end{vmatrix}, \quad \begin{vmatrix} a_1 & y_1 \\ a_2 & y_2 \end{vmatrix}, \quad \begin{vmatrix} x_1 & b_1 \\ x_2 & b_2 \end{vmatrix}, \quad \begin{vmatrix} x_1 & y_1 \\ x_2 & y_2 \end{vmatrix}.$$

This is most easily proved by expanding the first determinant and then picking out the terms in ab, in ay, in xb, and in xy. The reader should do this; it may help him to see just what happens in the proof of Theorem 23.

3. The product of two determinants

3.1. THEOREM 24. *If*

$$\Delta = \begin{vmatrix} a_1 & b_1 & c_1 \\ a_2 & b_2 & c_2 \\ a_3 & b_3 & c_3 \end{vmatrix}, \qquad \Delta' = \begin{vmatrix} x_1 & y_1 & z_1 \\ x_2 & y_2 & z_2 \\ x_3 & y_3 & z_3 \end{vmatrix},$$

the product $\Delta\Delta'$ *is equal to the determinant*

$$D = \begin{vmatrix} a_1x_1+b_1y_1+c_1z_1 & a_1x_2+b_1y_2+c_1z_2 & a_1x_3+b_1y_3+c_1z_3 \\ a_2x_1+b_2y_1+c_2z_1 & a_2x_2+b_2y_2+c_2z_2 & a_2x_3+b_2y_3+c_2z_3 \\ a_3x_1+b_3y_1+c_3z_1 & a_3x_2+b_3y_2+c_3z_2 & a_3x_3+b_3y_3+c_3z_3 \end{vmatrix}.$$

PROOF. By Theorem 23, Corollary 1, D is the sum of 27 determinants. These fall into two types, which we shall consider separately in (i) and (ii) below.

(i) There are 21 determinants† which have the same letter in two or more columns; each of these is equal to zero: e.g.

$$\begin{vmatrix} a_1x_1 & a_1x_2 & b_1y_3 \\ a_2x_1 & a_2x_2 & b_2y_3 \\ a_3x_1 & a_3x_2 & b_3y_3 \end{vmatrix} = x_1x_2y_3 \begin{vmatrix} a_1 & a_1 & b_1 \\ a_2 & a_2 & b_2 \\ a_3 & a_3 & b_3 \end{vmatrix} = 0,$$

the value of the last determinant being zero because two columns are the same.

(ii) There are 6 determinants with different letters a, b, c in

† The total 21 is made up thus: 3 have the same letter (a, b, or c) in each column; 6 have a in two columns with b or c in the third; 6 have b in two columns with c or a in the third; 6 have c in two columns with a or b in the third.

the different columns. These correspond to the 6 ways of arranging the letters a, b, c in different orders.

Consider any one of these orders, say a, c, b. The corresponding determinant is

$$\begin{vmatrix} a_1x_1 & c_1z_2 & b_1y_3 \\ a_2x_1 & c_2z_2 & b_2y_3 \\ a_3x_1 & c_3z_2 & b_3y_3 \end{vmatrix} = x_1y_3z_2 \begin{vmatrix} a_1 & c_1 & b_1 \\ a_2 & c_2 & b_2 \\ a_3 & c_3 & b_3 \end{vmatrix}$$

$$= -x_1y_3z_2 \begin{vmatrix} a_1 & b_1 & c_1 \\ a_2 & b_2 & c_2 \\ a_3 & b_3 & c_3 \end{vmatrix},$$

the minus sign appearing when we interchange columns in order to establish the alphabetical order a, b, c in the determinant on the right. We notice that the original order a, c, b, with a in 1st place, b in 3rd place, c in 2nd place, corresponds to the suffixes 1, 3, 2 for x, y, z. Moreover, $-x_1y_3z_2$ is a term in the expansion of Δ', the minus sign being given by the rule of § 1.

Consider a second order of the letters, say c, a, b. The corresponding determinant is

$$\begin{vmatrix} c_1z_1 & a_1x_2 & b_1y_3 \\ c_2z_1 & a_2x_2 & b_2y_3 \\ c_3z_1 & a_3x_2 & b_3y_3 \end{vmatrix} = +x_2y_3z_1 \begin{vmatrix} a_1 & b_1 & c_1 \\ a_2 & b_2 & c_2 \\ a_3 & b_3 & c_3 \end{vmatrix},$$

the plus sign appearing because we have to make *two* interchanges of columns in order to change the order c, a, b into the order a, b, c. Again, the order c, a, b, with a in 2nd place, b in 3rd place, c in 1st place, corresponds to the order 2, 3, 1 in the suffixes for x, y, z. Moreover, $+x_2y_3z_1$ is a term in the expansion of Δ'.

The results are similar for the other determinants of type (ii); each furnishes Δ multiplied by a term in the expansion of Δ'; and the 6 terms of this type furnish the 6 terms in the expansion of Δ'.

Hence $\qquad D = \Delta \times \sum \pm \{xyz\}_{123} = \Delta\Delta'.$

3.2. *Note on the proof of Theorem* 24. This proof becomes simpler when one knows how to handle a determinant of order n and knows the general rule for assigning + or − to the terms in the expansion of a determinant. The present proof is worth mastering, in order to convince oneself of the truth of the theorem; but the proof is not worth memorizing.†

The theorem and its immediate consequences are widely used.

4. Determinants whose elements are cofactors

In this section $A_1, A_2,..., C_3$ denote the cofactors of $a_1, a_2,..., c_3$ in the determinant

$$\Delta = \begin{vmatrix} a_1 & b_1 & c_1 \\ a_2 & b_2 & c_2 \\ a_3 & b_3 & c_3 \end{vmatrix}.$$

4.1. THEOREM 25.

$$\begin{vmatrix} A_1 & B_1 & C_1 \\ A_2 & B_2 & C_2 \\ A_3 & B_3 & C_3 \end{vmatrix} = \Delta^2.$$

PROOF. First suppose that $\Delta \neq 0$. Then

$$\begin{vmatrix} a_1 & b_1 & c_1 \\ a_2 & b_2 & c_2 \\ a_3 & b_3 & c_3 \end{vmatrix} \times \begin{vmatrix} A_1 & B_1 & C_1 \\ A_2 & B_2 & C_2 \\ A_3 & B_3 & C_3 \end{vmatrix}$$

$$= \begin{vmatrix} a_1 A_1+b_1 B_1+c_1 C_1 & a_1 A_2+b_1 B_2+c_1 C_2 & a_1 A_3+b_1 B_3+c_1 C_3 \\ a_2 A_1+b_2 B_1+c_2 C_1 & a_2 A_2+b_2 B_2+c_2 C_2 & a_2 A_3+b_2 B_3+c_2 C_3 \\ a_3 A_1+b_3 B_1+c_3 C_1 & a_3 A_2+b_3 B_2+c_3 C_2 & a_3 A_3+b_3 B_3+c_3 C_3 \end{vmatrix}.$$

By the results of Chapter XIII, §§ 6.3 and 6.4, this reduces to

$$\begin{vmatrix} \Delta & 0 & 0 \\ 0 & \Delta & 0 \\ 0 & 0 & \Delta \end{vmatrix} = \Delta^3.$$

Hence, when $\Delta \neq 0$,

$$\Delta \times \begin{vmatrix} A_1 & B_1 & C_1 \\ A_2 & B_2 & C_2 \\ A_3 & B_3 & C_3 \end{vmatrix} = \Delta^3,$$

† I wish that the proof of 'the product of determinants' were expressly excluded from all examinations below university examinations, when it should be included for n and not merely for 3.

and so

$$\begin{vmatrix} A_1 & B_1 & C_1 \\ A_2 & B_2 & C_2 \\ A_3 & B_3 & C_3 \end{vmatrix} = \Delta^2. \tag{1}$$

Now each side of the equation (1) is a polynomial in the variables a_1, a_2,..., c_3. We have proved the result subject only to the condition that the polynomial

$$\Delta = a_1 b_2 c_3 - a_1 b_3 c_2 + \dots \quad \neq 0. \tag{2}$$

Hence, by Theorem 7 (the 'irrelevance of polynomial inequalities'), (1) is true for all values of the variables a_1, a_2,..., c_3 and not merely for those that satisfy the inequality (2).

4.2. THEOREM 26.

$$\begin{vmatrix} B_2 & C_2 \\ B_3 & C_3 \end{vmatrix} = a_1 \Delta, \quad \begin{vmatrix} A_1 & C_1 \\ A_3 & C_3 \end{vmatrix} = b_2 \Delta, \quad -\begin{vmatrix} B_1 & C_1 \\ B_3 & C_3 \end{vmatrix} = a_2 \Delta,$$

the expressions on the left being respectively the cofactors of A_1, B_2, A_2 *in the determinant of Theorem* 25; *and so for the other cofactors.*

PROOF. We shall prove the first and the third of these results: the method of proof applies to all results of this type.

First suppose that $\Delta \neq 0$. Then

$$\begin{vmatrix} a_1 & b_1 & c_1 \\ a_2 & b_2 & c_2 \\ a_3 & b_3 & c_3 \end{vmatrix} \times \begin{vmatrix} 1 & 0 & 0 \\ A_2 & B_2 & C_2 \\ A_3 & B_3 & C_3 \end{vmatrix}$$

$$= \begin{vmatrix} a_1 & a_1 A_2 + b_1 B_2 + c_1 C_2 & a_1 A_3 + b_1 B_3 + c_1 C_3 \\ a_2 & a_2 A_2 + b_2 B_2 + c_2 C_2 & a_2 A_3 + b_2 B_3 + c_2 C_3 \\ a_3 & a_3 A_2 + b_3 B_2 + c_3 C_2 & a_3 A_3 + b_3 B_3 + c_3 C_3 \end{vmatrix}$$

$$= \begin{vmatrix} a_1 & 0 & 0 \\ a_2 & \Delta & 0 \\ a_3 & 0 & \Delta \end{vmatrix} = a_1 \Delta^2,$$

so that

$$\begin{vmatrix} B_2 & C_2 \\ B_3 & C_3 \end{vmatrix} = a_1 \Delta^2 / \Delta = a_1 \Delta.$$

Again

$$\begin{vmatrix} a_1 & b_1 & c_1 \\ a_2 & b_2 & c_2 \\ a_3 & b_3 & c_3 \end{vmatrix} \times \begin{vmatrix} A_1 & B_1 & C_1 \\ 1 & 0 & 0 \\ A_3 & B_3 & C_3 \end{vmatrix}$$

$$= \begin{vmatrix} a_1 A_1+b_1 B_1+c_1 C_1 & a_1 & a_1 A_3+b_1 B_3+c_1 C_3 \\ a_2 A_1+b_2 B_1+c_2 C_1 & a_2 & a_2 A_3+b_2 B_3+c_2 C_3 \\ a_3 A_1+b_3 B_1+c_3 C_1 & a_3 & a_3 A_3+b_3 B_3+c_3 C_3 \end{vmatrix}$$

$$= \begin{vmatrix} \Delta & a_1 & 0 \\ 0 & a_2 & 0 \\ 0 & a_3 & \Delta \end{vmatrix} = a_2 \Delta^2,$$

so that

$$-\begin{vmatrix} B_1 & C_1 \\ B_3 & C_3 \end{vmatrix} = a_2 \Delta.$$

This proves the results when $\Delta \neq 0$. But the latter is a polynomial inequality and is therefore irrelevant to the equality of two polynomials in the elements $a_1,\ldots, c_3$ (Theorem 7). Hence the results are true both when $\Delta = 0$ and when $\Delta \neq 0$.

4.3. *Special cases of Theorem* 26. When A, H, G,... denote the cofactors of a, h, g,... in

$$\Delta = \begin{vmatrix} a & h & g \\ h & b & f \\ g & f & c \end{vmatrix},$$

Theorem 26 gives the useful facts

$$BC-F^2 = a\Delta, \qquad GH-AF = f\Delta,$$
$$CA-G^2 = b\Delta, \qquad HF-BG = g\Delta,$$
$$AB-H^2 = c\Delta, \qquad FG-CH = h\Delta.$$

5. The cofactors of a zero determinant

5.1. THEOREM 27. *When* $\Delta = 0$,

$$B_1 C_2 = B_2 C_1, \qquad C_2 A_3 = C_3 A_2,$$

and so on; moreover, when the cofactors are not zero, the cofactors of any one row are proportional to the cofactors of any other row (and so for columns).

PROOF. By Theorem 26, the cofactor of any element A_1, $A_2,\ldots, C_3$ in

$$\begin{vmatrix} A_1 & B_1 & C_1 \\ A_2 & B_2 & C_2 \\ A_3 & B_3 & C_3 \end{vmatrix}$$

is equal to zero when $\Delta = 0$. Hence $B_2 C_3 - B_3 C_2 = 0$, $B_1 C_2 - B_2 C_1 = 0$, and so on. This proves the first part of the theorem.

We now wish to prove that A_1, B_1, C_1 are proportional to A_2, B_2, C_2. Suppose that at least one of the three numbers A_2, B_2, C_2 is not zero; say, $B_2 \neq 0$. Then, by what we have proved already,

$$A_1 B_2 = B_1 A_2, \qquad C_1 B_2 = C_2 B_1,$$

so that

$$A_1 = \frac{B_1}{B_2} A_2, \qquad B_1 = \frac{B_1}{B_2} B_2, \qquad C_1 = \frac{B_1}{B_2} C_2.$$

If, further, $B_1 \neq 0$, there is a number $k \neq 0$, such that

$$A_1 = kA_2, \qquad B_1 = kB_2, \qquad C_1 = kC_2;$$

that is, A_1, B_1, C_1 are proportional to A_2, B_2, C_2.

5.2. *Note on Theorem* 27. When some of the cofactors $A_1,\ldots, C_3$ are zero, care is necessary in applying the idea of one row of cofactors being proportional to another row. It is safest to go back to the forms $B_1 C_2 = B_2 C_1$, etc., and see what they yield.

As an example of what happens when zero cofactors occur, consider

$$\Delta = \begin{vmatrix} 1 & 2 & 3 \\ 4 & 5 & 6 \\ 8 & 10 & 12 \end{vmatrix} = 0,$$

in which

$$\begin{aligned} A_1 &= 0, & B_1 &= 0, & C_1 &= 0, \\ A_2 &= 6, & B_2 &= -12, & C_2 &= 6, \\ A_3 &= -3, & B_3 &= 6, & C_3 &= -3. \end{aligned}$$

The columns are proportional; the second and third rows are proportional.

The reader may prove for himself the result, '*if* $\Delta = 0$ *and if not more than one of* A_1, A_2, A_3 *is zero, and not more than one of* $B_1\ B_2\ B_3$ *is zero, then* B_1, B_2, B_3 *are proportional to* A_1, A_2, A_3'. There is, of course, a similar result concerning rows.

6. Other methods of multiplying determinants

The method of multiplying determinants given in Theorem 24 is commonly called 'multiplication by rows'; the process of forming the product is based on rows. Since determinants are unaltered when rows and columns are interchanged, there are other methods of forming the product. The two most important are

(1) 'multiplication by columns', when we work exactly as in Theorem 24 but use columns instead of rows;

(2) 'multiplication of row by column', when we use the rows of the first and the columns of the second; this is called 'matrix' multiplication since it is the rule for multiplying matrices.

We mention these methods in case the reader should encounter them in his reading of other books. In the examples which follow all multiplication is by rows.

7. Determinants of order four

The determinant

$$\begin{vmatrix} a_1 & b_1 & c_1 & d_1 \\ a_2 & b_2 & c_2 & d_2 \\ a_3 & b_3 & c_3 & d_3 \\ a_4 & b_4 & c_4 & d_4 \end{vmatrix}$$

is of order four; it has four columns and four rows. It denotes

$$a_1 \begin{vmatrix} b_2 & c_2 & d_2 \\ b_3 & c_3 & d_3 \\ b_4 & c_4 & d_4 \end{vmatrix} - b_1 \begin{vmatrix} a_2 & c_2 & d_2 \\ a_3 & c_3 & d_3 \\ a_4 & c_4 & d_4 \end{vmatrix} +$$

$$+ c_1 \begin{vmatrix} a_2 & b_2 & d_2 \\ a_3 & b_3 & d_3 \\ a_4 & b_4 & d_4 \end{vmatrix} - d_1 \begin{vmatrix} a_2 & b_2 & c_2 \\ a_3 & b_3 & c_3 \\ a_4 & b_4 & c_4 \end{vmatrix},$$

which can be further expanded into a form

$$\sum \pm (a_r b_s c_t d_u),$$

where r, s, t, u are the 24 permutations of the numbers 1, 2, 3, 4 and the plus or minus sign is fixed by a definite rule.

The properties II–VI, already proved for determinants of orders two and three, can be established for determinants of order four or more; some of the proofs are easy, some are hard. All such proofs can be deferred until the reader studies determinants of order n.

We give two examples of the evaluation of determinants of order four.

EXAMPLE 1. *Evaluate the determinant*

$$\Delta = \begin{vmatrix} 14 & 3 & 8 & 2 \\ 16 & 5 & 6 & 3 \\ 9 & 2 & 1 & 4 \\ 8 & 3 & 1 & 5 \end{vmatrix}.$$

SOLUTION. We aim, by adding multiples of columns, to get two or more zeros in the new first column [this to reduce the subsequent calculations].

$$\Delta = \begin{vmatrix} 0 & 3 & 8 & 2 \\ 0 & 5 & 6 & 3 \\ 4 & 2 & 1 & 4 \\ 1 & 3 & 1 & 5 \end{vmatrix} \qquad c_1' = c_1 - 2c_2 - c_3$$

$$= 4\begin{vmatrix} 3 & 8 & 2 \\ 5 & 6 & 3 \\ 3 & 1 & 5 \end{vmatrix} - \begin{vmatrix} 3 & 8 & 2 \\ 5 & 6 & 3 \\ 2 & 1 & 4 \end{vmatrix},$$

on expanding by the first column.† Thus

$$\Delta = 4\{3.27-5.38+3.12\}-\{3.21-5.30+2.12\}$$
$$= 4(117-190)-(87-150) = -229.$$

† The definition given expands by the first row, but the expansion can be regrouped (as with determinants of order three) to give expansions by any row or column.

EXAMPLE 2. *Prove that*

$$\begin{vmatrix} 1 & 1 & 1 & 1 \\ 1 & r & r^2 & r^3 \\ 1 & r^2 & r^3 & r^4 \\ 1 & r^3 & r^4 & r^5 \end{vmatrix} = 0.$$

SOLUTION (i). The expansion by the first row is the sum of four terms each of which is zero (the reader should write down the four terms like

$$\begin{vmatrix} r & r^2 & r^3 \\ r^2 & r^3 & r^4 \\ r^3 & r^4 & r^5 \end{vmatrix} - \begin{vmatrix} 1 & r^2 & r^3 \\ 1 & r^3 & r^4 \\ 1 & r^4 & r^5 \end{vmatrix} + \ldots$$

and see that each is zero because two columns are proportional).

SOLUTION (ii). Form a new determinant with

$$c_2' = c_2 - c_1, \qquad c_3' = c_3 - c_2, \qquad c_4' = c_4 - c_3$$

['columns that come before']; the result is

$$\begin{vmatrix} 1 & 0 & 0 & 0 \\ 1 & r-1 & r^2-r & r^3-r^2 \\ 1 & r^2-1 & r^3-r^2 & r^4-r^3 \\ 1 & r^3-1 & r^4-r^3 & r^5-r^4 \end{vmatrix} = \begin{vmatrix} r-1 & r^2-r & r^3-r^2 \\ r^2-1 & r^3-r^2 & r^4-r^3 \\ r^3-1 & r^4-r^3 & r^5-r^4 \end{vmatrix},$$

which is zero because the third column is r times the second.

8. Factors of determinants

The remainder theorem, coupled with the method of equating coefficients in polynomials known to be identical, often enables us to factorize a determinant with a minimum of calculation. Sometimes the addition of rows or columns will indicate factors of a determinant. The following are typical examples:

EXAMPLE 1. *Prove that*

$$\Delta \equiv \begin{vmatrix} a^2 & b^2 & c^2 \\ a & b & c \\ 1 & 1 & 1 \end{vmatrix} = (a-b)(a-c)(b-c).$$

SOLUTION. On expanding† Δ, we see that it may be regarded

(i) as a homogeneous polynomial in a, b, c of degree 3 in these variables;

† Rather, on thinking what the expansion would be if we worked it out.

(ii) as a non-homogeneous polynomial in a, of degree 2, whose coefficients are functions of b and c.

Also, $\Delta = 0$ when $a = b$, since it then has two columns the same. Hence, by the Remainder Theorem applied to a polynomial in a, Δ has a factor $a-b$.

Similarly, $b-c$ and $a-c$ are factors of Δ, and

$$\Delta \equiv K(a-b)(a-c)(b-c),$$

where K denotes the other factors of Δ, if any. Moreover, both Δ and the product $(a-b)(a-c)(b-c)$ are of degree 3 in the variables a, b, c. Hence K must be independent of a, b, c and so is a numerical constant (Theorem 6, Corollary 3). The coefficient of a^2b in the expansion of Δ is 1 and therefore $K = 1$ (Theorem 6, Corollary 2).

EXAMPLE 2. *Prove that*

$$\Delta \equiv \begin{vmatrix} a & b & c \\ c & a & b \\ b & c & a \end{vmatrix} = (a+b+c)(a+b\omega+c\omega^2)(a+b\omega^2+c\omega),$$

where ω is a complex cube root of unity.

SOLUTION (i).

$$\Delta = \begin{vmatrix} a+b+c & b & c \\ a+b+c & a & b \\ a+b+c & c & a \end{vmatrix} \qquad (c_1' = c_1+c_2+c_3),$$

and therefore Δ has a factor $a+b+c$.

Again,
$$c+a\omega+b\omega^2 = c\omega^3+a\omega+b\omega^2 \qquad (\omega^3 = 1)$$
$$= \omega(a+b\omega+c\omega^2),$$

and
$$b+c\omega+a\omega^2 = b\omega^3+c\omega^4+a\omega^2$$
$$= \omega^2(a+b\omega+c\omega^2),$$

so that

$$\Delta = \begin{vmatrix} a+b\omega+c\omega^2 & b & c \\ \omega(a+b\omega+c\omega^2) & a & b \\ \omega^2(a+b\omega+c\omega^2) & c & a \end{vmatrix} \qquad (c_1' = c_1+\omega c_2+\omega^2 c_3),$$

and therefore Δ has a factor $a+b\omega+c\omega^2$.

Similarly, Δ has a factor $a+b\omega^2+c\omega$ and so

$$\Delta \equiv K(a+b+c)(a+b\omega+c\omega^2)(a+b\omega^2+c\omega).$$

In this identity Δ and the product of the three linear factors are both of degree 3 in a, b, c. Hence K must be independent of a, b, c and so is a numerical constant. Since the coefficient of a^3 in the expansion of Δ is 1, K must be 1.

SOLUTION (ii). Consider the product by rows of the two determinants

$$\begin{vmatrix} a & b & c \\ c & a & b \\ b & c & a \end{vmatrix}, \quad \begin{vmatrix} 1 & 1 & 1 \\ 1 & \omega & \omega^2 \\ 1 & \omega^2 & \omega \end{vmatrix}.$$

It is, on making use of equations like

$$c+a\omega+b\omega^2 = c\omega^3+a\omega+b\omega^2 = \omega(a+b\omega+c\omega^2),$$

$$\begin{vmatrix} a+b+c & a+b\omega+c\omega^2 & a+b\omega^2+c\omega \\ a+b+c & \omega(a+b\omega+c\omega^2) & \omega^2(a+b\omega^2+c\omega) \\ a+b+c & \omega^2(a+b\omega+c\omega^2) & \omega(a+b\omega^2+c\omega) \end{vmatrix}$$

$$= (a+b+c)(a+b\omega+c\omega^2)(a+b\omega^2+c\omega)\begin{vmatrix} 1 & 1 & 1 \\ 1 & \omega & \omega^2 \\ 1 & \omega^2 & \omega \end{vmatrix},$$

and the result follows, since the determinant last written is equal to $3(\omega^2-\omega) \neq 0$. (This proof would, of course, be worthless if the last determinant were equal to zero.)

EXAMPLE 3. *Find the factors of the determinant*

$$\Delta \equiv \begin{vmatrix} 1 & 1 & 1 \\ a^2 & b^2 & c^2 \\ a^3 & b^3 & c^3 \end{vmatrix}.$$

SOLUTION. By the Remainder Theorem, $b-c$, $c-a$, $a-b$ are factors. The determinant is of degree 5 in a, b, c and so the remaining factor is of degree 2 in a, b, c. We shall give two methods of finding this factor, the first method being one of wide application, the second more particular in scope.

Method 1. [*Make the known factors appear explicitly.*]

$$\Delta \equiv \begin{vmatrix} 1 & 0 & 0 \\ a^2 & b+a & c+a \\ a^3 & b^2+ab+a^2 & c^2+ca+a^2 \end{vmatrix} (b-a)(c-a),$$

[Take $c_2' = c_2 - c_1$ and take out the factor $b-a$; $c_3' = c_3 - c_1$ and take out the factor $c-a$.]

$$\therefore \quad \Delta \equiv (b-a)(c-a)\begin{vmatrix} b+a & c+a \\ b^2+ab+a^2 & c^2+ca+a^2 \end{vmatrix}$$

$$\equiv (b-a)(c-a)(b-c)\begin{vmatrix} 1 & c+a \\ b+c+a & c^2+ca+a^2 \end{vmatrix}$$

[$c_1' = c_1 - c_2$ and take out the factor $b-c$]

$$\equiv (b-a)(c-a)(b-c)\begin{vmatrix} 1 & c+a \\ b+a & a^2 \end{vmatrix} \quad (r_2' = r_2 - c\,r_1)$$

$$\equiv (b-c)(c-a)(a-b)(bc+ca+ab).$$

Method 2. By the Remainder Theorem and the degree in a, b, c,

$$\Delta \equiv Q(b-c)(c-a)(a-b),$$

where Q is a quadratic in a, b, c.

Now when b and c are interchanged, both Δ and the product $(b-c)(c-a)(a-b)$ are multiplied by -1, and so Q is unaltered. Hence Q is symmetrical in b and c. Similarly, it is symmetrical in the three variables a, b, c. The only symmetrical quadratic function in a, b, c is

$$h(a^2+b^2+c^2)+k(bc+ca+ab),$$

where h, k are numerical constants.

But Δ contains no term in a^4; hence $h = 0$, and so

$$\begin{vmatrix} 1 & 1 & 1 \\ a^2 & b^2 & c^2 \\ a^3 & b^3 & c^3 \end{vmatrix} = k(bc+ca+ab)(b-c)(c-a)(a-b).$$

We see that $k = 1$, EITHER by considering the coefficient of b^2c^3 OR by giving a, b, c particular values.

9. Differentiation

9.1. THEOREM 28. *If the elements of the determinant*

$$\Delta \equiv \begin{vmatrix} a_1 & b_1 & c_1 \\ a_2 & b_2 & c_2 \\ a_3 & b_3 & c_3 \end{vmatrix}$$

are functions of x,

$$\frac{d\Delta}{dx} \equiv \begin{vmatrix} da_1/dx & b_1 & c_1 \\ da_2/dx & b_2 & c_2 \\ da_3/dx & b_3 & c_3 \end{vmatrix} + \begin{vmatrix} a_1 & db_1/dx & c_1 \\ a_2 & db_2/dx & c_2 \\ a_3 & db_3/dx & c_3 \end{vmatrix} + \begin{vmatrix} a_1 & b_1 & dc_1/dx \\ a_2 & b_2 & dc_2/dx \\ a_3 & b_3 & dc_3/dx \end{vmatrix}.$$

PROOF. In the notation of § 1,

$$\Delta = \sum \pm \{abc\}_{123},$$

and so

$$\frac{d\Delta}{dx} = \sum \pm \frac{d}{dx}\{abc\}_{123}.$$

But

$$\frac{d}{dx}(abc) = \frac{da}{dx}bc + a\frac{db}{dx}c + ab\frac{dc}{dx},$$

so that, when we put in the suffixes and collect together all terms in which an a is differentiated, we get a determinant whose first column is $\frac{da_1}{dx}, \frac{da_2}{dx}, \frac{da_3}{dx}$, second column b_1, b_2, b_3, and third column c_1, c_2, c_3; and so for terms in which a b or a c is differentiated.

COROLLARY. *There is a corresponding form for $d\Delta/dx$ in which we differentiate the rows of Δ instead of the columns.*

9.2. *Examples of differentiation.*

EXAMPLE 1. *Differentiate*

$$\Delta \equiv \begin{vmatrix} 1 & 1 & 1 \\ 1 & 2x & 3x^2 \\ x & x^2 & x^3 \end{vmatrix}$$

with respect to x.

SOLUTION. We shall differentiate 'by rows'.

$$\frac{d\Delta}{dx} = \begin{vmatrix} 0 & 0 & 0 \\ 1 & 2x & 3x^2 \\ x & x^2 & x^3 \end{vmatrix} + \begin{vmatrix} 1 & 1 & 1 \\ 0 & 2 & 6x \\ x & x^2 & x^3 \end{vmatrix} + \begin{vmatrix} 1 & 1 & 1 \\ 1 & 2x & 3x^2 \\ 1 & 2x & 3x^2 \end{vmatrix}.$$

The first and last of these determinants are equal to zero and

$$\frac{d\Delta}{dx} = x\begin{vmatrix} 1 & 1 & 1 \\ 0 & 2 & 6x \\ 1 & x & x^2 \end{vmatrix} = x\begin{vmatrix} 1 & 1 & 1 \\ 0 & 2 & 6x \\ 0 & x-1 & x^2-1 \end{vmatrix} \quad (r_3' = r_3 - r_1)$$

$$= 2x(x-1)\begin{vmatrix} 1 & 3x \\ 1 & x+1 \end{vmatrix} = 2x(x-1)(1-2x).$$

EXAMPLE 2. *Differentiate*

$$\Delta \equiv \begin{vmatrix} 1 & x & a \\ 1 & x^2 & a^2 \\ 1 & x^3 & a^3 \end{vmatrix}$$

with respect to x.

SOLUTION. We shall differentiate 'by columns'; there will be only one non-zero determinant as only one column of Δ contains x.

$$\frac{d\Delta}{dx} = \begin{vmatrix} 1 & 1 & a \\ 1 & 2x & a^2 \\ 1 & 3x^2 & a^3 \end{vmatrix} = -\begin{vmatrix} 1 & 1 & a \\ 2x & 1 & a^2 \\ 3x^2 & 1 & a^3 \end{vmatrix}$$

$$= -a^2(a-1)+2xa(a^2-1)-3x^2a(a-1)$$

$$= -a(a-1)\{a-2x(a+1)+3x^2\}.$$

EXAMPLES XIV

1. Prove that

$$\begin{vmatrix} a-x & a-y \\ b-x & b-y \end{vmatrix} = (a-b)(x-y).$$

2. Prove that

$$\begin{vmatrix} a-x & a-y & a-z \\ b-x & b-y & b-z \\ c-x & c-y & c-z \end{vmatrix} = 0.$$

3. By considering the product of the two determinants

$$\begin{vmatrix} a^2 & a & 1 \\ b^2 & b & 1 \\ c^2 & c & 1 \end{vmatrix}, \quad \begin{vmatrix} 1 & -2x & x^2 \\ 1 & -2y & y^2 \\ 1 & -2z & z^2 \end{vmatrix},$$

prove that

$$\Delta \equiv \begin{vmatrix} (a-x)^2 & (a-y)^2 & (a-z)^2 \\ (b-x)^2 & (b-y)^2 & (b-z)^2 \\ (c-x)^2 & (c-y)^2 & (c-z)^2 \end{vmatrix}$$

$$= 2(b-c)(c-a)(a-b)(y-z)(z-x)(x-y).$$

4. Obtain the result of Example 3 by direct consideration of the determinant Δ.

5. Prove that

$$\begin{vmatrix} (a-x)^2 & (a-y)^2 & (a-z)^2 & (a-w)^2 \\ (b-x)^2 & (b-y)^2 & (b-z)^2 & (b-w)^2 \\ (c-x)^2 & (c-y)^2 & (c-z)^2 & (c-w)^2 \\ (d-x)^2 & (d-y)^2 & (d-z)^2 & (d-w)^2 \end{vmatrix} = 0.$$

HINT. *Either* multiply

$$\begin{vmatrix} a^2 & a & 1 & 0 \\ b^2 & b & 1 & 0 \\ c^2 & c & 1 & 0 \\ d^2 & d & 1 & 0 \end{vmatrix}$$

by an appropriate determinant in x, y, z, w, *or* use the ideas of § 2; write down one of the 81 ($= 3^4$) determinants indicated by Theorem 23 and convince yourself that it and all the others must be zero.

6.** By looking at Examples 1 and 3 guess the value of the determinant obtained by writing $(a-x)^3$, etc., instead of $(a-x)^2$, etc., in Example 5.

Verify your guess.

7.** By looking at Examples 2 and 5 write down a determinant of order five whose first row is

$$(a-x)^3 \quad (a-y)^3 \quad (a-z)^3 \quad (a-u)^3 \quad (a-v)^3$$

and whose value is zero.

Verify that the value is zero.

---

8. Prove that

$$\begin{vmatrix} 0 & a-b \\ b-a & 0 \end{vmatrix} = (a-b)^2, \qquad \begin{vmatrix} 0 & a-b & a-c \\ b-a & 0 & b-c \\ c-a & c-b & 0 \end{vmatrix} = 0.$$

9. Prove that

$$\begin{vmatrix} 0 & y-x & 0 \\ x-y & 0 & a-b \\ 0 & b-a & 0 \end{vmatrix} = 0, \qquad \begin{vmatrix} 0 & f & g \\ -f & 0 & h \\ -g & -h & 0 \end{vmatrix} = 0.$$

10. Prove that

$$\begin{vmatrix} 0 & a & 0 & x \\ -a & 0 & -x & 0 \\ 0 & x & 0 & a \\ -x & 0 & -a & 0 \end{vmatrix} = (a^2-x^2)^2.$$

Prove the results 11–14:

11. $$\begin{vmatrix} a & b & c \\ a^2 & b^2 & c^2 \\ b+c & c+a & a+b \end{vmatrix} = (a+b+c)(b-c)(c-a)(a-b).$$

12. $$\begin{vmatrix} 1 & 1 & 1 \\ a & b & c \\ bc & ca & ab \end{vmatrix} = (b-c)(c-a)(a-b).$$

13. $$\begin{vmatrix} 1 & 1 & 1 \\ a & b & c \\ a^3 & b^3 & c^3 \end{vmatrix} = (b-c)(c-a)(a-b)(a+b+c).$$

14. $$\begin{vmatrix} 1 & 1 & 1 \\ a^2 & b^2 & c^2 \\ a^3 & b^3 & c^3 \end{vmatrix} = (b-c)(c-a)(a-b)(bc+ca+ab).$$

15. Prove the results of Examples 13 and 14 by expressing

$$\Delta \equiv \begin{vmatrix} 1 & 1 & 1 & 1 \\ x & a & b & c \\ x^2 & a^2 & b^2 & c^2 \\ x^3 & a^3 & b^3 & c^3 \end{vmatrix}$$

as the product of factors and then considering

(i) the cofactors of x^2 and x in Δ,

(ii) the coefficients of x^2 and x in the product.

In the following examples s_r denotes $a^r+b^r+c^r$ and σ_r denotes a^r+b^r.

16. Prove that

$$\begin{vmatrix} s_0 & s_1 & s_2 \\ s_1 & s_2 & s_3 \\ s_2 & s_3 & s_4 \end{vmatrix} = \begin{vmatrix} 1 & 1 & 1 \\ a & b & c \\ a^2 & b^2 & c^2 \end{vmatrix}^2.$$

17. Prove that

$$\begin{vmatrix} 1 & 1 & 1 \\ a & b & c \\ a^3 & b^3 & c^3 \end{vmatrix} \times \begin{vmatrix} 1 & 1 & 1 \\ a & b & c \\ a^2 & b^2 & c^2 \end{vmatrix} = \begin{vmatrix} s_0 & s_1 & s_2 \\ s_1 & s_2 & s_3 \\ s_3 & s_4 & s_5 \end{vmatrix}.$$

18.** Find the factors of

$$\begin{vmatrix} s_0 & s_1 & s_2 \\ s_2 & s_3 & s_4 \\ s_4 & s_5 & s_6 \end{vmatrix}, \quad \begin{vmatrix} s_0 & s_2 & s_4 \\ s_2 & s_4 & s_6 \\ s_4 & s_6 & s_8 \end{vmatrix}.$$

HINT. Study the suffixes and indices in Example 17.

19. Prove that

$$\begin{vmatrix} \sigma_0 & \sigma_1 & \sigma_2 \\ \sigma_1 & \sigma_2 & \sigma_3 \\ 1 & x & x^2 \end{vmatrix} = \begin{vmatrix} 1 & 1 & 0 \\ a & b & 0 \\ 0 & 0 & 1 \end{vmatrix} \times \begin{vmatrix} 1 & 1 & 1 \\ a & b & x \\ a^2 & b^2 & x^2 \end{vmatrix}$$

$$= (a-b)^2(x-a)(x-b).$$

20.** Find the factors of

$$\begin{vmatrix} s_0 & s_1 & s_2 & s_3 \\ s_1 & s_2 & s_3 & s_4 \\ s_2 & s_3 & s_4 & s_5 \\ 1 & x & x^2 & x^3 \end{vmatrix}.$$

HINT. Study Example 19.

In Examples 21–4 the symbols Δ, A_1, B_1,... have the meanings assigned to them in § 4.

21. Prove that

$$\Delta_1 \equiv \begin{vmatrix} a_1+x & b_1 & c_1 \\ a_2 & b_2+x & c_2 \\ a_3 & b_3 & c_3+x \end{vmatrix}$$

$$\equiv x^3+(a_1+b_2+c_3)x^2+(A_1+B_2+C_3)x+\Delta.$$

22. Prove that

$$\Delta_2 \equiv \begin{vmatrix} A_1+y & B_1 & C_1 \\ A_2 & B_2+y & C_2 \\ A_3 & B_3 & C_3+y \end{vmatrix}$$

$$\equiv y^3+(A_1+B_2+C_3)y^2+(a_1+b_2+c_3)\Delta y+\Delta^2.$$

23. Prove that, if $\Delta \neq 0$ and x_1, x_2, x_3 are the roots of the equation $\Delta_1 = 0$ (Example 21), then the roots of the equation $\Delta_2 = 0$ (Example 22) are Δ/x_1, Δ/x_2, Δ/x_3.

24. Prove that the vertices of the triangle formed by the three lines whose equations are

$$a_1x+b_1y+c_1 = 0, \quad a_2x+b_2y+c_2 = 0, \quad a_3x+b_3y+c_3 = 0,$$

are $(A_1/C_1, B_1/C_1)$, $(A_2/C_2, B_2/C_2)$, $(A_3/C_3, B_3/C_3)$,

and show that the area of the triangle is $\Delta^2/2C_1C_2C_3$.

25. When $\Delta \equiv abc+2fgh-af^2-bg^2-ch^2 = 0,$
$A = bc-f^2,$
$F = gh-af,$ etc.,

prove that $A(Ax^2+2Hxy+By^2+2Gx+2Fy+C) = (Ax+Hy+G)^2.$

---

26. Prove that, when u, v, w are functions of x and dashes denote differentiations with regard to x,

$$\frac{d}{dx}\begin{vmatrix} u & v & w \\ u' & v' & w' \\ u'' & v'' & w'' \end{vmatrix} = \begin{vmatrix} u & v & w \\ u' & v' & w' \\ u''' & v''' & w''' \end{vmatrix}.$$

27.† Prove that

$$\frac{\partial^2}{\partial x \partial y}\begin{vmatrix} 1 & 1 & 1 \\ x & y & z \\ x^2 & y^2 & z^2 \end{vmatrix} = 2\begin{vmatrix} 1 & 1 \\ x & y \end{vmatrix}; \qquad \frac{\partial^3}{\partial x \partial y \partial z}\begin{vmatrix} 1 & 1 & 1 \\ x & y & z \\ x^2 & y^2 & z^2 \end{vmatrix} = 0.$$

28.† Prove that, when

$$\Delta \equiv \begin{vmatrix} 1 & 1 & 1 \\ x & y & z \\ x^2 & y^2 & z^2 \end{vmatrix},$$

$$\frac{\partial \Delta}{\partial x}+\frac{\partial \Delta}{\partial y}+\frac{\partial \Delta}{\partial z} = 0.$$

HINT. In spite of appearances differentiate by rows and not by columns; and, if you can, think of the *operator* $\frac{\partial}{\partial x}+\frac{\partial}{\partial y}+\frac{\partial}{\partial z}$ as a differential operator whose effect on a determinant is given by a theorem just like Theorem 28.

29. The elements of a determinant are functions of x and three rows become equal when $x = a$. Prove that the differential coefficient of the determinant with regard to x contains $x-a$ as a factor.

30. Any polynomial of degree n in x which contains $x-a$ as a factor may be written in the form

$$p_0(x-a)+p_1(x-a)^2+\ldots+p_n(x-a)^n.$$

Use this fact to show that if $F(x)$ is a polynomial, and if both $F(x)$ and $F'(x)$ contain $x-a$ as a factor, then $F(x)$ must contain $(x-a)^2$ as a factor.

Use the result of Example 29 to prove that, if three rows of a determinant become equal when $x = a$, the determinant contains $(x-a)^2$ as a factor.

† Examples 27 and 28 should be omitted by readers who have not yet learnt 'partial differentiation'.

ANSWERS

Examples I a

1. Linear, rational, cubic, rational, rational.

Examples I b

1. $-28, -\frac{1}{2}, -\frac{25}{22}$. 2. $550, 10, 96\frac{1}{4}$. 3. (i) 8.

Examples II a

1. (i) $f(-1) = -3-7+6 = -4$;
   (ii) $f(2) = 80+56+5 = 141$;
   (iii) $f(-3) = -27+9-3+1 = -20$;
   (iv) $f(-2) = -8+4+2+1 = -1$.

2. (i) $a = -2$; (ii) $a = 8$.

3. (i) $a = -3, b = 4$; (ii) $a = 7, b = -1$.

7. $a = 36, b = 2$.

8. (i) $x+1, x-2, 2x-1$; (ii) $x-2, 3x+1, 2x+3$.

9. (i) $x-1, x-2, x-3$; (ii) $x-1, x-2, x+3$; (iii) $x-1, x+1, x-2, x+3$.

10. (i) $x-a, x-2a, x-3a$; (ii) $x-y, x-2y, x+3y$.

11. (i) $a+b, a-2b, 2a-b$; (ii) $b-2c, 3b+c, 2b+3c$.

Examples II b

1. (i) $a = 1, b = -4, c = 4$; (ii) $a = 1, b = -3, c = 3, d = -1$; (iii) $a = 3, b = 11, c = 14, d = 5$.

4. (i) $a = 1, b = 1$; (ii) $a = 1, b = 1, c = 1$; (iii) $a = 1, b = 3, c = 9, d = 2$.

5. (i) $a = 1, b = -6, c = 0, d = -2$; (ii) $a = 1, b = -2, c = -1$; (iii) $a = 1, b = -1, c = -3, d = -11$.

Examples II c

1. $b = -3a, c = 3a, d = -a$. 3. $5, -15/4$.

4. $a = 1, b = 2, c = 4$.

5. $(x+y+z)(2x-3y+4z)$, $(\frac{1}{2}x+y+z)(4x-3y+4z)$.

EXAMPLES III A

1. $5, -25, 50.$ 2. $13, 799.$

3. $x^2-80x+27=0$; $y^2-11y+27=0$; $3y^2-5y+1=0$;
$y^2-35y+303=0.$

5. (i) $cx^2+bx+a=0$; (ii) $ay^2+y(b-2a)+a-b+c=0$;
(iii) $ay^2+2by+4c=0$; (iv) $a^2x^2-(b^2-2ac)x+c^2=0$;
(v) $a^3x^2+b(b^2-3ac)x+c^3=0$; (vi) $a^3x^2+abcx+c^3=0.$

11. $-9, -5, 5/3, 19.$

12. (i) $7x^3+4x^2-2x+1=0$; (ii) $y^3-5y^2+11y=0$;
(iii) $y^3-4y^2+8y-15=0$; (iv) $y^3+4y^2+44y-49=0.$

EXAMPLES III B

5. (i) $\frac{1}{2}(7\pm\sqrt{37})$; (ii) $\frac{1}{2}(7\pm i\sqrt{3})$; (iii) $\frac{1}{2}(5\pm\sqrt{17})$; (iv) $\frac{1}{2}(5\pm i\sqrt{3})$.

6. (i) $x^2-4x+1=0$; (ii) $x^2-4x+13=0$; (iii) $x^2-6x+4=0$;
(iv) $x^2-6x+14=0$; (v) $4x^2-4x-1=0$; (vi) $x^2-x+1=0.$

EXAMPLES V A

1. Turning-points are $(-\frac{3}{4}, -\frac{1}{8})$; $(-\frac{1}{3}, \frac{2}{3})$; $(1, -5)$; all minima.

2. Turning-points are $(-\frac{1}{3}, 3\frac{1}{3})$; $(-1, 5)$; $(\frac{1}{2}, 6\frac{1}{4})$; all maxima.

3. (i) $(-1, 14)$ max., $(2, -13)$ min. (ii) $(-1, -14)$ min., $(2, 13)$ max.
(iii) $(-2, -9)$ min., $(2, 23)$ max. (iv) $(-2, 9)$ max., $(2, -23)$ min.

4. $k=0, -1$; the graph has max. at $(1, 0)$ and min. at $(2, -1)$.

5. Inflexion at $(\frac{1}{2}, -\frac{1}{2})$; tangent parallel to Ox.

6. Inflexion at $(2, -1)$; tangent parallel to Ox.

7. Inflexion at $(0, 9)$; slope of tangent 12.

8. (i) Min. at $(1, 0)$; max. at $(2, 1)$; min. at $(3, 0)$.
(ii) Max. at $(-1, 11)$; min. at $(1, 3)$.
(iii) Inflexion at $(1, 20)$; tangent parallel to Ox;

$$\frac{dy}{dx}=15(x-1)^2(x^2-2x+3).$$

9. Max. at $x \simeq -2$; min. at $x \simeq \frac{1}{3}$.

EXAMPLES V B

1. (i) 1·3; (ii) 1·3; (iii) 2.

2. (i) 0·4; (ii) 0·4; (iii) 0·5.

3. $-1·1$, 0·8, 1·5.

4. 0·9 (0·857), $-2·5$.

EXAMPLES VI A

5. (i) 120; (ii) 720. 6. $2^3.3.7.13.17.19 = 705{,}432$.

7. (i) $6\times 9 = 54$; (ii) $15\times 84 = 1{,}260$.

8. (i) 120; (ii) 15; (iii) 56; (iv) 56.

9. $7! = 5040$, $4! = 24$, $5! = 120$.

10. $10.9.8.7.6.5.4 = 604{,}800$. 11. 604,800.

12. 120. 13. 60, 2520, 10, 21.

16. (i) $35+35 = 70$; (ii) $20+15 = 35$; (iii) $28+112+70 = 210$.

EXAMPLES VI B

1. $2^4+4.2^3.x+6.2^2.x^2+4.2.x^3+x^4 = 16+32x+24x^2+8x^3+x^4$,
$3^5+5.3^4(2x)+10.3^3(2x)^2+10.3^2(2x)^3+5.3(2x)^4+(2x)^5$
$= 243+810x+1080x^2+720x^3+240x^4+32x^5$,
$1+6x+15x^2+20x^3+15x^4+6x^5+x^6$,
$1+5(2x)+10(2x)^2+10(2x)^3+5(2x)^4+(2x)^5$
$= 1+10x+40x^2+80x^3+80x^4+32x^5$.

2. (i) $x^7+7x^6+21x^5+35x^4+35x^3+21x^2+7x+1$,
(ii) $x^5+5.2x^4+10.2^2x^3+10.2^3x^2+5.2^4x+2^5$
$= x^5+10x^4+40x^3+80x^2+80x+32$,
(iii) $(2x)^3+3(2x)^2.3+3(2x).3^2+3^3 = 8x^3+36x^2+54x+27$,
(iv) $x^8+8x^7+28x^6+56x^5+70x^4+56x^3+28x^2+8x+1$.

3. 16, 3072, 36. 4. 45, 84, 80.

5. $3+17x+41x^2+55x^3+45x^4+23x^5+7x^6+x^7$,
$x+5x^2+10x^3+10x^4+5x^5+x^6$.

6. -5, 40. 7. 1·149, 2·558.

8. 1·0161, 1·1566. 9. 0·9039, 0·9378.

10. $x^7+7x^5+21x^3+35x+35x^{-1}+21x^{-3}+7x^{-5}+x^{-7}$; -14, 0.

11. $x^4+4x^2+6+4x^{-2}+x^{-4}$, $x^6+7x^4+19x^2+26+19x^{-2}+7x^{-4}+x^{-6}$.

12. $8+12x-30x^2-35x^3+45x^4+27x^5-27x^6$; -16×217; -3^7 (expand in descending powers of x).

13. $a_4 = 81$, $a_3 = 108$, $a_2 = 378$, $a_1 = 336$, $a_0 = 595$. 14. 29.

15. $1-10x+25x^2$, $1+24x+240x^2$, $27+27x-18x^2$.

16. $6(a-b)$, $3(5a^2-12ab+5b^2)$. 17. $\frac{1}{8}$.

19. ${}_nC_r\,4^{n-r}3^rx^r$, $(-1)^r{}_nC_r\,5^{n-r}2^rx^r$, ${}_nC_r\,2^ra^{n-r}b^r$.

20. ${}_nC_r\,2^{n-r}3^rx^{2n-3r}$, ${}_nC_r\,2^{n-r}3^rx^{n-3r}$.

EXAMPLES VI c

1. 70, 252. 5. 945/64. 6. $r = 11$.

9. ${}_{n+3}C_r = {}_nC_r + 3\,{}_nC_{r-1} + 3\,{}_nC_{r-2} + {}_nC_{r-3}$,

$${}_{n+4}C_r = {}_nC_r + 4\,{}_nC_{r-1} + 6\,{}_nC_{r-2} + 4\,{}_nC_{r-3} + {}_nC_{r-4}.$$

12. Coeff. of x^r in $(1+x)^m(1-x)^{m+1}$ = coeff. of x^r in $(1-x)(1-x^2)^m$.

13. Coeff. of x^{n+r} in $(1+x)^{2n}$ = coeff. of x^{n+r} in $(1+x)^n . x^n(1+x^{-1})^n$.

14. Coeff. of x^n in $(1+x)^{2n+m}$ = coeff. of x^n in $(1+x)^n x^{m+n}(1+x^{-1})^{m+n}$.

15. Coeff. of x^n in $(1+x)^{2n+m}$ = coeff. of x^n in $(1+x)^{n+m} x^n(1+x^{-1})^n$.

16. L.H.S. $= (1+x)^n +$ the $S(x)$ of Problem 3, § 7.5.

17. Let L.H.S. $= f(x)$, and $F(x) = \int f(x)\,dx$; then $x^{-1}F(x) =$ the $S(x)$ of Problem 3, § 7.5.

19. Let $f(x) = 1.2 + 2.3\,{}_nC_1 x + \ldots$; let $\int f(x)\,dx = F(x)$, $\int F(x)\,dx = G(x)$; then $f(x)$ is the second differential coefficient of $G(x)$, and

$$G(x) = x^2(1+x)^n.$$

20. *Second part.* On integrating $F''(x) = (1+x)^n$ twice $[F'(0) = 0,\ F(0) = 0]$,

$$F(x) = \frac{(1+x)^{n+2}}{(n+1)(n+2)} - \frac{x}{n+1} - \frac{1}{(n+1)(n+2)}.$$

EXAMPLES VII

1. (i) Homog. of degree 3, (ii) and (iii) not homogeneous, (iv) homog. of degree 2; (v) homog. of degree 3, (vi) not homogeneous.

3. (i), (iv), and (v) are symmetrical;

   (ii) is unaltered if we interchange x and y, when it becomes $xz+zy+2yx$, which is the same as $yz+zx+2xy$; but when we interchange x and z we get $yx+xz+2zy$, which is not the same as $yz+zx+2xy$;

   (iii) is altered if we interchange x and z.

5. (i) $bc(b-c)+ca(c-a)+ab(a-b)$; (ii) $a^2(b-c)+b^2(c-a)+c^2(a-b)$;

   (iii) $b^3-c^3+c^3-a^3+a^3-b^3$; (iv) $bc(b^2+c^2)+ca(c^2+a^2)+ab(a^2+b^2)$;

   (v) $a^2b+b^2c+c^2a$; (vi) $a(b-c)+b(c-a)+c(a-b)$.

6. (i) $x^4+y^4+z^4$; (ii) $xy^2+yz^2+zx^2$;

   (iii) $yz(y-z)+zx(z-x)+xy(x-y)$; (iv) $y^2+z^2+z^2+x^2+x^2+y^2$.

Examples VIII a

1. (i) $A = \frac{7}{2}$, $B = -16$, $C = \frac{31}{2}$; (ii) $A = -1$, $B = 3$, $C = 5$.
(iii) $A = 1$, $B = -\frac{9}{5}$, $C = \frac{2}{5}$, $D = \frac{1}{5}$.

2. (i) $\dfrac{1}{x-1}+\dfrac{2}{x+1}-\dfrac{1}{2x-1}$, (ii) $\dfrac{1}{x-1}-\dfrac{2}{x-2}+\dfrac{1}{x-3}$,
(iii) $\dfrac{1}{x+1}-\dfrac{2}{x+2}+\dfrac{1}{x+3}$, (iv) $\dfrac{1}{x-3}-\dfrac{2}{x-4}+\dfrac{1}{x-5}$.

3. (i) $\dfrac{1}{4(x-1)}+\dfrac{5}{7(2x-1)}+\dfrac{25}{28(x+3)}$,
(ii) $-\dfrac{9}{10(x-1)}+\dfrac{21}{55(3x+2)}+\dfrac{17}{22(x-3)}$,
(iii) $-\dfrac{5}{4(x-2)}-\dfrac{7}{36(x+2)}+\dfrac{26}{9(2x-5)}$,
(iv) $\dfrac{8}{x-2}-\dfrac{31}{x-3}+\dfrac{26}{x-4}$.

4. (i) $\dfrac{2}{x-1}-\dfrac{2x+1}{x^2+1}$, (ii) $\dfrac{1}{x-2}+\dfrac{x+1}{x^2+x+5}$,
(iii) $-\dfrac{1}{8(2x-1)}+\dfrac{7(2x+1)}{8(4x^2+9)}$, (iv) $-\dfrac{5}{2(x-1)}+\dfrac{3}{x-2}+\dfrac{x+1}{2(x^2+1)}$.

5. (i) $A = -1$, $B = -1$, $C = 1$; (ii) $A = 1$, $B = -1$, $C = 2$;
(iii) $A = -1$, $B = -1$, $C = -1$, $D = 1$;
(iv) $A = 3$, $B = 1$, $C = -3$, $D = 2$.

6. (i) $-\dfrac{1}{x-2}-\dfrac{1}{(x-2)^2}+\dfrac{1}{x-3}$, (ii) $\dfrac{1}{x}-\dfrac{1}{x-1}+\dfrac{2}{(x-1)^2}$,
(iii) $-\dfrac{1}{x-2}-\dfrac{1}{(x-2)^2}-\dfrac{1}{(x-2)^3}+\dfrac{1}{x-3}$,
(iv) $\dfrac{3}{x-1}+\dfrac{1}{(x-1)^2}-\dfrac{3}{x-2}+\dfrac{2}{(x-2)^2}$.

7. (i) $\dfrac{2}{x-2}+\dfrac{3}{x-3}+\dfrac{4}{x-4}$, (ii) $\dfrac{2}{x}-\dfrac{3}{x-1}+\dfrac{1}{x-2}$,
(iii) $\dfrac{2}{x-1}+\dfrac{3}{(x-1)^2}+\dfrac{1}{x+1}$, (iv) $\dfrac{2}{2x-1}-\dfrac{x}{x^2+1}$,
(v) $\dfrac{1}{9(x-2)}-\dfrac{2x-5}{9(2x^2+1)}$, (vi) $\dfrac{1}{x-1}+\dfrac{2}{(x-1)^2}+\dfrac{1}{(x-1)^3}-\dfrac{2}{2x-1}$.

8. (i) Is obtained from the given result by replacing x by $x+1$,
(ii) „ „ „ „ x by $x-1$.

9. (i) $\frac{1}{2}\left(\frac{1}{x-1}-\frac{1}{x+1}\right)$, (ii) $\frac{1}{x-1}+\frac{1}{x+1}$.

10. $1+\frac{2}{x-1}+\frac{3}{x-2}$, $1-\frac{2}{x-1}+\frac{3}{x-2}$.

Examples VIII B

1. Use $(N_1+D_1)/(N_1-D_1)$. 2. Use $(N_1+3D_1)/(2N_1-D_1)$.

3. 2, 22/7. 4. $2:3:4$. 5. $1:2:3$.

6. $1:3:2$. 7. $x = 35t$, $y = -4t$, $z = 10t$.

8. $x = 2$, $y = 3$, $z = 4$.

9. $x = -1$, $y = 1$, $z = 3$; $x = 1$, $y = -1$, $z = -3$.

Examples IX A

1. Asymptotes (i) $y = 1$; (ii) $y = -1$: general shape § 1.4.

2. Asymptotes (i) $y = 0$; (ii) $y = 0$: general shape § 2.2.

3. (i) Asymptote $y = 0$; has max. at $(1, \frac{1}{2})$, min. at $(-1, -\frac{1}{2})$.
(ii) Asymptote $y = 1$; has max. at $(1, \frac{3}{2})$, min. at $(-1, \frac{1}{2})$.

4. Asymptotes (i) $x = 2$, $y = 1$; (ii) $x = \frac{3}{2}$, $y = 1$: general shape § 2.3.

5. Asymptotes (i) $x = 1$, $x = 3$, $y = 0$; (ii) $x = 1$, $x = -1$, $y = 0$: no max., no min.

6. (i) Asymptotes $x = 1$, $x = 2$, $y = 0$; turning-points $x = 3\pm\sqrt{2}$.
(ii) Asymptotes $x = 1$, $x = 2$, $y = 0$; turning-points $x = \pm\sqrt{2}$.

7. (i) Asymptote $y = 1$; general shape § 2.2 inverted.
(ii) Asymptote $y = \frac{1}{2}$; general shape § 3.2, with max. at $x \simeq -0{\cdot}37$ and min. at $x \simeq 1{\cdot}37$.

Examples X

1. (i) $1+3x+6x^2+10x^3+\ldots+\frac{1}{2}(r+1)(r+2)x^r+\ldots$.
(ii) $1+\frac{1}{2}x-\frac{1}{8}x^2+\frac{1}{16}x^3-\ldots+(-1)^{r-1}\frac{1.3\ldots(2r-3)}{2.4\ldots 2r}x^r+\ldots$.

2. (i) $-17\frac{1}{2}$, $-\frac{7.9.11}{8}$; $(-1)^k\frac{3.5\ldots(2k+1)}{1.2\ldots k}$; $-\frac{1}{2} < x < \frac{1}{2}$.
(ii) $-4\frac{2}{3}$, $-30\frac{1}{3}$; $(-1)^k.\frac{1.4.7\ldots(3k-2)}{1.2.3\ldots k}$; $-\frac{1}{3} < x < \frac{1}{3}$.

3. (i) r; (ii) 1.

4. (i) $1+2x+2x^2+\ldots+2x^r+\ldots$.
(ii) $2x+5x^2+9x^3+\ldots+\frac{1}{2}r(r+3)x^r+\ldots$.

5. (i) $\frac{n(n-1)\ldots(n-r+1)}{r!}2^{n-r}$, (ii) $\frac{n(n-1)\ldots(n-r+1)}{r!}2^{n-r}3^r$,

(iii) $(-1)^r\frac{n(n-1)\ldots(n-r+1)}{r!}3^{n-r}2^r$.

Valid (i) $-2 < x < 2$, (ii) $-\frac{2}{3} < x < \frac{2}{3}$, (iii) $-\frac{3}{2} < x < \frac{3}{2}$.

6. (i) $\frac{1}{4}-\frac{1}{4}x+\frac{3}{16}x^2-\ldots+(-1)^r\frac{r+1}{2^{r+2}}x^r+\ldots$; $-2 < x < 2$.

(ii) $\frac{1}{\sqrt{2}}\left\{1+\frac{1}{4}x+\frac{3}{32}x^2+\ldots+\frac{1.3\ldots(2r-1)}{2.4\ldots 2r}\frac{x^r}{2^r}+\ldots\right\}$; $-2 < x < 2$.

(iii) $\frac{1}{27}+\frac{1}{27}x+\frac{2}{81}x^2+\ldots+\frac{1}{2}(r+1)(r+2)\frac{x^r}{3^{r+3}}+\ldots$; $-3 < x < 3$.

7. (i) $1+3x+7x^2+\ldots+(2^{r+1}-1)x^r+\ldots$; $-\frac{1}{2} < x < \frac{1}{2}$.
(ii) $1-3x+7x^2-\ldots+(-1)^r(2^{r+1}-1)x^r+\ldots$; $-\frac{1}{2} < x < \frac{1}{2}$.

8. (i) $2+3x+5x^2+\ldots+(2^r+1)x^r+\ldots$; $-\frac{1}{2} < x < \frac{1}{2}$.
(ii) $2+3x+4x^2+\ldots+(r+2)x^r+\ldots$; $-1 < x < 1$.

9. (i) $1+3x+7x^2+\ldots+(r^2+r+1)x^r+\ldots$; $-1 < x < 1$.
(ii) $x^2+4x^3+11x^4+\ldots+(2^r-r-1)x^r+\ldots$; $-\frac{1}{2} < x < \frac{1}{2}$.

10. (i) 0·914(3); by tables, 0·915.
(ii) 0·929; by tables, 0·929.

11. (i) $1{\cdot}006\times 10^{-6}$; same by tables.
(ii) 999·9; by tables, 1000.

12. (i) $1+4x+12x^2+36x^3$, (ii) $1+2x+x^2-4x^3$.

13. (i) $1-5x+10x^2-10x^3$, (ii) $1-3x^2+10x^3$.

14. (i) $\frac{1}{2^8}(2-21x+126x^2-567x^3)$, (ii) $\frac{1}{2^5}\left(1-\frac{5x}{4}+\frac{15x^2}{32}+\frac{35x^3}{128}\right)$.

16. $a = 2$, $b = 4$. 17. $a = 18$, $b = -20$, $c = 8$.

18. First prove that $a = 1$, $b = -2$, $c = 1$.

Examples XI a

1. $3x+\frac{1}{2}x^2+x^3+\frac{1}{4}x^4$; $-1 < x < 1$.

2. $1+x+\frac{x^2}{2!}+\ldots+\frac{x^n}{n!}+\ldots$,

$1-2x+\frac{4x^2}{2!}-\ldots+(-1)^n\frac{2^nx^n}{n!}+\ldots$,

$1-3x+\frac{9x^2}{2!}-\ldots+(-1)^n\frac{3^nx^n}{n!}+\ldots$.

5. (i) Coeff. of $x^n = 0$, (ii) coeff. of $x^n = 2(n+1)/n!$.

9. (i) $-\frac{5}{6}x-\frac{25}{72}x^2-...-\frac{5^n x^n}{6^n n}-...;\ -\frac{6}{5} \leqslant x < \frac{6}{5}$.

   (ii) $(2^{-n}+3^{-n})/n;\ -2 \leqslant x < 2$.

10. $\log 2+\frac{3}{2}x+\frac{3}{8}x^2+...+\frac{1}{n}\left\{1+\frac{(-1)^{n-1}}{2^n}\right\}x^n+...;\ -1 \leqslant x < 1$.

11. $4x-3x^2+\frac{4}{3}x^3;\ 5x-\frac{13}{2}x^2+\frac{35}{3}x^3$.

12. $x+\frac{1}{2}x^2-\frac{2}{3}x^3+\frac{1}{4}x^4$.

Examples XI B

1. $a = \frac{2}{3}$, $b = \frac{1}{6}$.

3. $\frac{1}{2}\log\frac{1+x}{1-x}$. 4. $\frac{1}{2x}\log\frac{1+x}{1-x}$.

Examples XIII A

1. 11, 66, -2, 43.

2. $xy(x-y)$, $bx-ay$, $(x-y)(1-x-y-xy)$.

3. 50, 51, 52, 53.

4. II, III. 5. IV, Corollary, II. 6. V, V.

7. VI, $r_2' = r_2-5r_1$; VI, $r_1' = r_1-2r_2$.

8. $\frac{x}{-1} = \frac{y}{2} = \frac{z}{-1}$.

9. $\frac{x}{-13} = \frac{y}{34} = \frac{z}{53}$.

10. $\frac{x}{-5} = \frac{y}{-4} = \frac{z}{9}$.

11. $\frac{x}{-51} = \frac{y}{22} = \frac{z}{13}$.

12. $\frac{x}{1} = \frac{y}{2} = \frac{z}{1}$.

13. $\frac{x}{-1} = \frac{y}{2} = \frac{z}{1}$.

14. $x = 1$, $y = -2$.

15. $x = -\frac{13}{53}$, $y = \frac{34}{53}$.

16. $x = -\frac{5}{9}$, $y = -\frac{4}{9}$.

17. $x = -\frac{51}{13}$, $y = \frac{22}{13}$.

18. $x = 1$, $y = 2$.

19. $x = -1$, $y = 2$.

Examples XIII B

1. -9, -2, 43.

2. $-(y-z)(z-x)(x-y)$ *or* $y^2z+z^2x+x^2y-yz^2-zx^2-xy^2$,
   $x^3+y^3+z^3-3xyz$.

3. $ax(y^2-z^2)+by(z^2-x^2)+cz(x^2-y^2)$. The second determinant has the same value as the first.

EXAMPLES XIII C

1. $x(mc-nb)-y(lc-na)+z(lb-ma)$, $x(bn-cm)-y(an-cl)+z(am-bl)$, $a(yn-zm)-b(xn-zl)+c(xm-yl)$, — The order *mc* or *cm* does not matter, but keep same pattern throughout.

   $x^2(y-z)-y^2(x-z)+z^2(x-y)$,

   $x^2(y-z)-x(y^2-z^2)+yz(y-z)$,

   $x^2(bn-cm)-a(y^2n-z^2m)+l(y^2c-z^2b)$.

2. $x(mc-nb)-l(yc-zb)+a(yn-zm)$,

   $x(bn-cm)-a(yn-zm)+l(yc-zb)$,

   $a(yn-zm)-x(bn-cm)+l(bz-cy)$,

   $x^2(y-z)-x(y^2-z^2)+yz(y-z)$,

   $x^2(y-z)-y^2(x-z)+z^2(x-y)$,

   $x^2(bn-cm)-y^2(an-cl)+z^2(am-bl)$.

3. (i) each $= \sum ayz(b-c)$,

   (ii) each $= \sum xa(b^2-c^2)$,

   (iii) each $= \sum xbc(c-b)$.

EXAMPLES XIII D

1. $-436, -7, 0$. 2. $0, -261, 984$.

EXAMPLES XIII E

1. (i) $X = vn-wm$, $Y = wl-un$, $Z = um-vl$.

   (ii) $X = cq-br$, $Y = ar-cp$, $Z = bp-aq$.

   (iii) $X = 1$, $Y = -2$, $Z = 1$.

4. $x = 1, y = 1, z = -1$. 5. $x = 4, y = 1, z = 0$.

6. $x = 1, y = 2, z = 3$. 7. $x = 2, y = 1, z = 2$.

8. $x = (b-d)(c-d)/(b-a)(c-a)$, etc.

13. $x = 11-10\lambda$, $y = -2+3\lambda$, $z = \lambda$; where λ is arbitrary.

14. $x = 4+4\lambda$, $y = 1+13\lambda$, $z = 5\lambda$; where λ is arbitrary.

15. $x = \frac{1}{2}(5-\lambda)$, $y = \frac{1}{2}(7-\lambda)$, $z = \lambda$; where λ is arbitrary.

Examples XIV

6. $9(a-b)(a-c)(a-d)(b-c)(b-d)(c-d)(x-y)(x-z)$ etc.

7. The first column is $(a-x)^3$, $(b-x)^3$, $(c-x)^3$, $(d-x)^3$, $(e-x)^3$. The proof consists of multiplying a determinant whose first row is

$$a^3 \quad a^2 \quad a \quad 1 \quad 0$$

by one whose first row is

$$1 \quad -3x \quad 3x^2 \quad x^3 \quad 0;$$

another proof by Theorem 24, when every possible determinant is zero.

18. (i) $(b^2-c^2)(c^2-a^2)(a^2-b^2)(b-c)(c-a)(a-b)$;

(ii) $\{(b^2-c^2)(c^2-a^2)(a^2-b^2)\}^2$.

20.
$$\begin{vmatrix} 1 & 1 & 1 & 0 \\ a & b & c & 0 \\ a^2 & b^2 & c^2 & 0 \\ 0 & 0 & 0 & 1 \end{vmatrix} \times \begin{vmatrix} 1 & 1 & 1 & 1 \\ a & b & c & x \\ a^2 & b^2 & c^2 & x^2 \\ a^3 & b^3 & c^3 & x^3 \end{vmatrix}$$
$$= \{(b-c)(c-a)(a-b)\}^2(x-a)(x-b)(x-c).$$

NOTATIONS

$f(x)$	a function of x
$=$	equals, is equal to
$\equiv$	is identically equal to
$\neq$	is not equal to
$\simeq$	is approximately equal to
i	$\sqrt{(-1)}$
ω	$\cos(2\pi/3)+i\sin(2\pi/3)$, $\omega^3 = 1$
$n!$	$n(n-1)(n-2)...1$
${}_nP_r$	$n!/(n-r)!$
${}_nC_r$	$n!/r!\,(n-r)!$
$>$	is greater than
$<$	is less than
$\sum$	denotes summation
Δ	denotes a determinant

INDEX

Numbers refer to pages.

PRINTED IN GREAT BRITAIN
AT THE UNIVERSITY PRESS, OXFORD
BY VIVIAN RIDLER
PRINTER TO THE UNIVERSITY